The Rhetoric of Project Apollo

The Rhetoric of Project Apollo

The Moon Story and Its Televised Memories

Kathy K. Previs

LEXINGTON BOOKS
Lanham • Boulder • New York • London

Published by Lexington Books
An imprint of The Rowman & Littlefield Publishing Group, Inc.
4501 Forbes Boulevard, Suite 200, Lanham, Maryland 20706
www.rowman.com

86-90 Paul Street, London EC2A 4NE, United Kingdom

British Library Cataloguing in Publication Information Available

Library of Congress Cataloging-in-Publication Data

Names: Previs, Kathy K., 1965– author.
Title: The rhetoric of Project Apollo : the moon story and its televised memories / Kathy K. Previs.
Description: Lanham : Lexington Books, 2025. | Includes bibliographical references and index. | Summary: "The Rhetoric of Project Apollo examines the metaphorical language that CBS News, with the assistance of NASA's Public Affairs Office, used in reporting the Apollo missions, 1968–1972, and explains how its cultural narratives have shifted in recent broadcast remembrances"—Provided by publisher.
Identifiers: LCCN 2024037327 (print) | LCCN 2024037328 (ebook) | ISBN 9781666974904 (cloth) | ISBN 9781666974911 (epub)
Subjects: LCSH: Project Apollo (U.S.)—Public relations. | Project Apollo (U.S.)—Press coverage. | United States. National Aeronautics and Space Administration—Public relations. | Television broadcasting of news—United States. | Mass media—United States—Influence.
Classification: LCC TL789.8.U6 A574 2025 (print) | LCC TL789.8.U6 (ebook) | DDC 629.45/4—dc23/eng/20240828
LC record available at https://lccn.loc.gov/2024037327
LC ebook record available at https://lccn.loc.gov/2024037328

∞™ The paper used in this publication meets the minimum requirements of American National Standard for Information Sciences—Permanence of Paper for Printed Library Materials, ANSI/NISO Z39.48-1992.

To the memory of my mother, Norma Rich Kerbaugh—the brightest of stars—whose light continues to shine.

Contents

Acknowledgments

This research would not have been possible without the assistance and support of many people and institutions. Because this book's journey began with a doctoral dissertation, my infinite appreciation to committee chair Joseph Slade, who offered invaluable guidance, and to committee members Katherine Jellison, Joe Bernt, and Don Flournoy, all of whom offered wonderful insights in their areas of specialty. My thanks to Ohio University's Contemporary History Institute, an excellent program that provided rigorous instruction and research funding in the manuscript's preliminary stages.

With the support of a Guggenheim Fellowship from the Smithsonian Institution's National Air and Space Museum, ideas for this topic were fine-tuned. Roger Launius, who rekindled my interest in Apollo, Margaret Weitekamp, and Valerie Neal provided information, insight, and enthusiasm and got me thinking about space and history in new ways. Many thanks also to Mike Neufeld, David DeVorkin, Martin Collins, Paul Ceruzzi, Cathy Lewis, Allen Needell, Teasel Muir-Harmony, and John Krige, who also offered excellent suggestions as my research progressed. Shout out to NASM's library staff, who were instrumental in locating specific pieces of information.

At the NASA History Office, special thanks to Nadine Andreassen, Steve Garber, John Hargenrader, Colin Fries, and the late Jane Odom, all of whom happily aided me in finding hard-to-find information throughout the years. I am grateful to each for their support and long-standing friendships. I am also very appreciative of long-time friends and wonderful former colleagues at NASA Headquarters: Tina Utz, who provided graphics support, and Sally Mee, who gave me a place to stay while I completed my fellowship.

I also owe a debt of gratitude to Joel Banow, Special Events Director at *CBS Evening News* during the Apollo years, and PAOs from NASA Kennedy Space Center for providing me with fascinating oral histories of their Apollo experiences. They, along with their coworkers, made the Apollo broadcasts possible, and I hope I have done them justice in recounting their stories. Sincere thanks also to Brad Perry of NASA Langley Research Center for sharing his inspiring NASA story and CBS News experience.

Huge hat tip to the staff at the Television News Archives at Vanderbilt University, especially to John Lynch and Skip Pfeiffer, whose humor made traveling through 50 years of space reporting and multiple hours with Walter Cronkite and Eric Sevareid go by much more quickly. Also in Nashville, I am much indebted to Michael Gray at the Country Music Hall of Fame for tracking down the tapes of songs that went to the Moon with Apollos 12 and 14. Thank you to David Eason, whose graduate courses piqued my interest in cultural studies and introduced me to influential scholars such as James Carey and Michael Schudson and whose own scholarship gave me an appreciation of music's cultural influence. To Nancy and Gary Koger, thank you for letting me stay so I could research and write during the early stages of this project.

My sincere appreciation to Nicolette Amstutz at Lexington/Bloomsbury Books, whose guidance has been remarkable. Thanks to Pam Parry for your encouragement and support, and to my colleagues in higher education, near and far—too many to name here but who are beyond appreciated. Special thanks to Debra Baida for allowing me to use the perfect photograph, for cover suggestions, and for your friendship.

Saving the best for last, heartfelt thanks and much love to my family, especially my parents, Norma and Larry, and siblings, Patty, Nancy, and Jay, who have always had my back and believed in me. Most of all, thank you to my husband and life partner, Ken Previs; I love and admire you. You each had a hand in making me who I am today.

Introduction

In May 1961, President John F. Kennedy announced to Congress in his now-famous "before this decade is out" speech that, "while we cannot guarantee that we shall one day be first, we can guarantee that any failure to make this effort will find us last."[1] It was important to beat the Soviets in the Space Race, for that would mean a defeat of Communist ideology. In January 2004, to garner support for the Mars Exploration Rovers, President George W. Bush announced,

> Mankind is drawn to the heavens for the same reason we were once drawn to unknown lands and across the open sea. We choose to explore space because doing so improves our lives and lifts our national spirit. . . . The vision I outline today is a journey, not a race.[2]

It seems that decades after the United States last put a man on the Moon when the Apollo missions ended in 1972, history was beginning to repeat itself, whether the metaphor used was "race" or "journey."

Since the Apollo Moon landing, the United States has had numerous successes in space, from Voyager's flybys of Saturn, Jupiter, Neptune, and Uranus in the 1980s, to the Hubble Space Telescope and the Mars Pathfinder, to the International Space Station in the 1990s. With Bush's 2004 announcement of a plan to establish a permanent presence on the Moon, Americans' memory of the Apollo program has resurfaced in the media and continues to do so with recent successes such as NASA's ARTEMIS mission in 2010 to orbit the Moon. Next was news of mission successes such as the first edible food being grown in space and the Pioneer and Voyager spacecraft exploring Saturn's atmosphere. Most recently, stories of Perseverance investigating Mars to determine whether it was ever habitable to microbial life while confirming various theories of weather events have rekindled interest in space-related accomplishments by NASA and Space Exploration Technologies Corporation (known as SpaceX), founded by Elon Musk to reduce space transportation costs and eventually to establish colonies on Mars. Similarly, reports of today's new era of space tourism funded by magnates Jeff Bezos and Richard Branson have recaptured the American imagination. In these representations of space exploration, how has the rhetoric of space exploration changed, and what do these changes imply? How have the language and the metaphors used to describe Project Apollo (then and now) as an American event shaped the nation's perception of the space program? This is important because, as scholars have long realized, the funding sources of a scientific or technological mission are often determined by the public's approval of that mission.[3] The ways in which space exploration is presented on television will be revealed by examining the language our culture uses to communicate space travel and discovery. Additionally, has this rhetoric changed over time, and if so, what do those changes tell us about a society that has historically professed a fascination with exploration?

More than 50 years of journalism about the Apollo Program have produced no scholarly analysis of how the Apollo Program has been presented to the public on television and how it continues to be broadcast. When television is mentioned in discussions of Apollo, journalists merely say that the program's events were among the most watched in the history of the United States. Moreover, it has been said that television contributed to the success of NASA and its programs as much as NASA and its programs perpetuated television's success in the 1960s, a time when both NASA and television technologies were in their youth.[4] Undoubtedly, this relationship—between medium and space spectacle—calls for further investigation.

As communication scholars have noted for decades, television has steadily become a primary source of entertainment and information for Americans, a trend that began in the late 1960s and early 1970s, just as the Apollo Project reached its zenith. In 1960, 87 percent of American homes were equipped with a television,[5] 96 percent by 1970[6]; by the mid-1970s, a Roper public opinion poll indicated that Americans relied more on television than newspapers for information.[7] On July 20, 1969, the historic day man set foot on the Moon, 94 percent of all homes with a television were tuned in,[8] excluding the number of people gathered in public places to witness the event. Clearly, most Americans learned of and followed the Apollo program via television. Thus, the findings included here are fresh because scholars have focused on print media coverage. Several factors have delayed the analysis of television's role: the time involved in analyzing hours of videotape, the expense related to viewing the footage in television archives, and the effort of finding footage that dates back more than 50 years—a time, of course, before VCRs entered the marketplace.

In addition to being a primary source of information, television entertains. Indeed, one of the debates in communication and history disciplines is the notion of the "glamorization" or "popularization" of history: at what point is television informational and at what point is it a dramatic medium? Of equal importance, how does the language used to describe events "glamorize" or "popularize" a narrative? While current literature focuses on the dichotomy between "fact" and "glamorization" of history for socio-political topics, a comparative analysis of scientific and/or technological subjects as represented on television (news and documentaries) has not been conducted for the U.S. space program. This book will explore the tensions between popular history and professional history as played out in the media arena of television, focusing on one area of the space program—Apollo. By examining the rhetoric used in the networks' framing of this scientific and technological mission, it will place it in its historical and cultural context during the Apollo mission years from 1968–1972, subsequent years from 1973 to the 2004 25th year anniversary of the historic Apollo 11, and continuing to 2022, the 50-year anniversary of Apollo 17, Apollo's last mission.

Using television as a medium to analyze Apollo coverage also seems to fit nicely with discursive communication. While historians tend to gauge documentaries and other made-for-television productions in terms of historical accuracy, communication researchers are more interested in how the train of language and images reflects social and political concerns. Here, the author is interested in the rhetorical variations used to describe Apollo by journalists (e.g., the evening news) and popular chroniclers (e.g., The History Channel). In short, this work will discover how media outlets such as CBS, CNN, The History Channel (THC), and PBS have portrayed the Apollo Program and its mostly positive consequences.

Scholars have narrowed the "popularization" debate to events such as the Vietnam War, World War II, the Holocaust, and the Civil War, for examples. While these events are of course important, representations of science and technology in cultural memory have been neglected by communication researchers. Here I will explore the ways that employing metaphors made the complexities of rocket science and space exploration understandable to the American public who funded the Apollo program. While a vast amount of research focuses on political rhetoric, only a handful of scholars have focused on how the shifting of metaphors has contributed to a better understanding of technology and science and their roles in society.[9] For example, metaphors changed as the goals of the Apollo project mutated: what began as a race into space evolved into space exploration for scientific discovery.

And today, discussions of Apollo have acted as a catalyst to either revisit something that was missing in the original narratives or provide justification to continue space exploration for the betterment of humankind. In this manuscript, I pinpoint when the rhetoric began to change and uncover the metaphors that are more currently used to represent Apollo now that we have had 50 years to understand the meaning of the Space Race. In sum, retrospection enables us to better understand news reports and documentaries of space exploration.

Although Apollo has increasingly been in the news, especially in comparison with nearly every new space venture, anniversaries of the missions, and each passing of its key players, not all stories have been favorable since Apollo's golden days. For example, the February 2003 Columbia space shuttle explosion that killed all seven astronauts on board has evoked the Apollo project in nostalgic reflex: Apollo seems to embody NASA's "glory years." Critics spoke of bringing back Apollo-style management to oversee shuttle management. Realistically, today's economy requires private-sector support for research and development from companies like SpaceX or consortiums with international space programs. Additionally, recent successful attempts by China to be the third country to participate in space exploration have revived discussions of the American Moon landing and what its political and scientific accomplishments were. For instance, China's Chang'e 4 mission was the first soft landing on the far side of the Moon on January 3, 2019. Finally, with the recent success of the Perseverance Mars Rover landing and news of more women working as astronauts and engineers, discussions of Apollo have resurfaced in theories of how space exploration ideally should work: inclusion of a more diverse workforce, with a renewed focus on reporting women's roles in Apollo's success 50 years later.

It is therefore necessary to step back in time to gauge several factors that influenced Apollo and its televised coverage. One must look beyond the political climate to place the reader in the proper historical context. For example, it is not enough just to say, "Apollo was created as a result of the Cold War," but rather, "How was the American public convinced billions of dollars should be spent to send man to the Moon?" This was especially important as Apollo was the largest project undertaken by the government during peacetime and perhaps the largest technical undertaking at any point in history,[10] costing American taxpayers $25 billion in 1960s dollars, more than $226 billion in today's dollars.[11]

In answering this question, I enlist what communication theorist James W. Carey calls symbolic representations that a culture shares and examine how these symbols are communicated through rhetoric. In addition to explaining the political climate of the day, a discussion of how the media portrayed it is provided. Chapter 1 explains Carey's ritual view of communication and offers examples of how it can be applied to space exploration and specifically Project Apollo. This discussion will be beneficial in understanding subsequent chapters.

To expand on Carey's insights, I utilize a rhetorical analysis methodology adapted from Celeste Condit's discussion of genetic discourse. Cultures employ a variety of metaphors to symbolize what they hold dear and what defines them as a culture, especially in narratives that seek to explain complex scientific and technological endeavors. Chapter 2 probes the importance of metaphors and explains why unlocking their rhetorical meanings is important in a communication-based narrative. It also provides a discussion detailing the utility of oral history as a method for conducting a project that is grounded in historical memory.

Rhetoric used to present the Apollo program to the public in the early years of space exploration originated from NASA and the three television news networks, particularly CBS. Chapter 3 examines the role of NASA's public relations efforts overseen by its Public Affairs Office. It indicates to what degree NASA produced its own metaphors as compared to those television used to explain Apollo's missions to the public. Was the American public spoon-fed government-generated information, or did network television devise its own scheme of transmitting information? Oral histories of several NASA Public Affairs Officers are included to determine the source of important, significant information regarding Apollo, as well as press releases and other public documents specific to each Apollo mission.

Following this same logic, then, I take a deep dive into how television portrayed the program. Just as NASA answered to the public, so did the networks with the additional considerations of advertisers and shareholders they had to appease. Like today, numerous factors influenced evening news content during the Apollo years. Chapter 4 utilizes the Shoemaker and Reese hierarchical model of organization to illustrate the ways in which media organizations produce content. A variety of influences determine content: individual media workers, media routines, structures of the organization, factors outside of the organization, and the organization's ideology.[12] This chapter deconstructs these five influences on the *CBS Evening News* as it presented the Apollo story during the late 1960s and early 1970s. CBS was selected instead of NBC and ABC because it had a concrete number one rating for network evening news beginning in the early 1960s,[13] and Walter Cronkite, its popular anchor, has been credited with solidifying these ratings due to his enthusiastic reporting of the Apollo missions. Finally, CBS News was by far the most innovative of the three networks in creating the technologies needed to broadcast man's first missions to the Moon. Therefore, focusing on CBS helps to understand the evolution of those technologies as Apollo progressed.

The discursive themes and narrative structures CBS used in stories pertaining to Apollo from 1968 to 1972 are presented in part II, chapters 5, 6, 7, and 8. Here, I am interested in how the metaphors that were used to discuss man's journey to the Moon changed over the course of Project Apollo. The answers to such an inquiry will provide clues as to how the network continued to present stories of science and technology even though public interest clearly faded with each successful mission. To further gauge change in such representations, I also analyze the narrative structures after the Apollo missions were complete, from 1973 to 2022. Analyses of the discursive themes and metaphors during this later period will help explain how the collective memory of Apollo has been defined through its televised representations on CBS, CNN, THC, and PBS. Just as each of the missions had a different objective, so do the networks. CBS News and CNN cable news inform; THC entertains; and the public broadcasting network, PBS, with its NOVA series, enlightens. Again, part of the goal of this research is to uncover how Apollo is remembered on television. Thus, the aim here is not to compare a news channel (e.g., CBS) with an entertainment channel (e.g., THC), but rather to discover how each venue, as a media institution, employs rhetoric to remember Project Apollo.

Chapter 5 examines nationalism and the extent to which the networks became not only advocates for space exploration but also conveyed the notion of American excellence and its "can do" culture. Chapter 6 addresses romanticism, or passion and spectacle, over that which is useful. Here, notions of exploration, heroism, frontierism, challenge, human spirit, adventure, drama, spirituality, peace, and curiosity all play significant roles in appealing to audiences in the United States and around the world. Chapter 7 unpacks the shift to pragmatism when the initial Apollo missions met Kennedy's challenge and public support began to wane. Similarly, chapter 8 highlights yet another rhetorical shift in technology discussions, specifically how the Apollo program used technology for military purposes but also to illustrate America's leadership in the tech arena and for its entertainment value. More importantly, chapter 8 also examines the importance of technology workers and how the long overdue focus on NASA's diverse workforce has provided new angles to the customary Apollo story while also paying tribute to the women who made the Moon missions successful.

Findings are summarized in chapter 9 and are applied to the larger social context. The presentation of science and technology matters because Americans want their country to conquer new frontiers yet quickly become apathetic with news of such stories. It is notable that some of the metaphors used to tell the Apollo story have endured for 50 years, such as the incorporation of purely American symbols and numerous examples of U.S. technological prowess, while others have fallen by the wayside. In a world where private investors have entered the realm of space technology and travel, rhetoric shifts yet again as the need to convince taxpayers of funding massive space endeavors, while still important, becomes less prevalent in the production of space news. As women's roles in space exploration replace those of office workers to engineer and astronaut, while simultaneously their roles in newsrooms have shifted from reporter to producer and anchor, mediated stories of one of the greatest accomplishments in U.S. history reflect the cultural changes in American life. In the end, the Apollo story mimics the human story of risk, adventure, perseverance, and change, and therefore continues to resonate with a worldwide audience.

NOTES

1. "Address to Joint Session of Congress, May 25, 1961," John F. Kennedy Presidential Library and Museum, accessed June 10, 2024, https://www.jfklibrary.org/learn/about-jfk/historic-speeches/address-to-joint-session-of-congress-may-25-1961.

2. "President Bush Announces New Vision for Space Exploration Program," The White House, accessed June 10, 2024. https://georgewbush-whitehouse.archives.gov/news/releases/2004/01/20040114-3.html.

3. Celeste Condit, *The Meaning of the Gene: Public Debates about Human Heredity* (Madison, WI: The University of Wisconsin Press, 1999), Donald MacKenzie, *Inventing Accuracy: A Historical Sociology of Nuclear Missile Guidance* (Cambridge, MA: MIT Press, 1990), and Dorothy Nelkin, *Selling Science: How the Press Covers Science and Technology* (New York: W. H. Freeman and Company, 1987).

4. Mary Ann Watson, *The Expanding Vista: American Television in the Kennedy Years* (New York: Oxford University Press, 1990).

5. Ibid., 8.

6. Joshua Meyrowitz, *No Sense of Place: The Impact of Electronic Media on Social Behavior* (New York: Oxford University Press, 1985), 133.

7. John E. O'Connor, ed., *American History, American Television: Interpreting the Video Past* (New York: Frederick Ungar Publishing Co., 1983), 380.

8. Ibid., 376.

9. See, for example, Celeste Condit and Dorothy Nelkin.

10. The History Channel, *Tech Effect: Apollo 11*, originally aired June 20, 2004.

11. Alex Roland, "Late night with Aaron Brown," *CNN News*, aired January 14, 2004. According to dollartimes.com, the original figure of $25 billion computes to $216 billion in 2024 dollars. The figure is adjusted for annual inflation, which over this period was 4 percent.

12. Pamela Shoemaker and Stephen Reese, *Mediating the Message: Theories of Influences on Mass Media Content,* 2nd ed. (White Plains, NY: Longman Publishers, 1991).

13. Garrick Utley, *You Should Have Been There Yesterday: A Life in Television News* (Collingdale, PA: Diane Publishing Co., 2000).

Part I

THE APOLLO MOON STORY: COMMUNICATION, NASA, AND TELEVISION

To understand how narrative structures were used to communicate complex scientific and technological missions to the Moon, it is first necessary to provide a background of multiple communication factors that contributed to the Apollo story. Chapter 1 provides an overview Project Apollo and the importance of analyzing television as a maker and keeper of myths that were created and are now maintained in telling the exciting Moon story. Because several considerations go into providing commentary to the public, a close look at symbolic representations using James Carey's ritual view of communication model to describe how viewers make sense of news and other mediated stories is offered. But first a cursory discussion of NASA's inception and a description of each Apollo mission and its purposes are provided to explain how the lunar expeditions were and continue to be presented in news and documentaries. The use of metaphors as a rhetorical strategy to appeal to audiences is discussed in chapter 2 in tandem with a narrative on the utility of oral history since the review of public relations tools such as press releases and government documents merely skims the surface of historical inquiry. Chapter 3 provides a brief history of NASA and takes an in-depth look at its Public Affairs Office during the Apollo years, the agency's public relations tactics, and its role in producing rhetoric in multiple forms that were shared with news agencies. Given time and other industry constraints, chapter 4 uncovers ways in which television produces content based on the Shoemaker and Reese hierarchical model of organization. These factors proved to be paramount in the never before attempted task of planning and executing live coverage of the missions to the Moon, as evident in discussions of CBS News anchor Walter Cronkite's contributions to Apollo coverage and special events director Joel Banow's technical innovations in covering the moonshots.

Chapter 1

Apollo, Mythmakers, and Ritual Communication

On October 4, 1957, the Soviet Union launched Sputnik, an artificial satellite that orbited the Earth. The American public was shocked that a communist nation traditionally lagging in technology could beat America at anything, especially an America that mass-produced nifty gadgets and all the latest appliances for the late-1950s consumer culture. Worse was the consensus that, if the Russians controlled space, they would also control the Earth and all its inhabitants. Part of the blame for coming in second in science and technology, said President Dwight D. Eisenhower, was attributable to declining math and science scores among American students across the country. Although he wanted to devise a plan that would make the United States competitive in exploration, Eisenhower insisted that there was no military or economic value in a "race" with the Soviet Union.[1] According to media historian Pam Parry, Eisenhower's authorization of NASA might have alleviated the Soviet's success with Sputnik, but there was still a public perception problem. "The president knew America still had superior satellite technology, but he could not tell the public without tipping his hand to the Soviets."[2]

Conversely, John F. Kennedy and Richard M. Nixon used the 1960 election campaign to emphasize the importance of the space race as a means to make the United States number one technologically, militarily, and ideologically in the eyes of the world. However, when John F. Kennedy became president in 1960, he was not fully concerned with the Soviet "threat" until Russian cosmonaut Yuri Gagarin became the first man to orbit the Earth. Not until after the Bay of Pigs fiasco did he approach his advisors and ask them to figure out what, if anything, the United States could do to regain credibility, to be viewed as a "leader" again, to revive the American spirit, and, most importantly, to beat the Russians.[3] Thus arose the myth that, if the United States conquered space, it would win the Cold War.

The response, of course, was the decision to send a human to the Moon. On May 25, 1961, in a "Special Message to the Congress on Urgent National Needs," President Kennedy announced, "I believe that this nation should commit itself to achieving the goal, before this decade is out, of landing a man on the Moon and returning him safely to Earth."[4] At the time Kennedy made this commitment, the United States had a grand total of 15 minutes of experience in space with Alan Shepard's suborbital flight. NASA engineers worked day and night devising plans that might or might not have worked—there were a lot of unknowns as neither man nor machine had ever been launched to the Moon and back. To test the physical laws of space, NASA created two projects, Mercury and Gemini, which would perfect the Moon mission. Mercury (1961–1963) and Gemini (1965–1966) would also provide television with news and entertainment because with each new mission came another "American first." As each successful mission brought the president's promise that much closer to reality, audiences around the world readied themselves for the climax of the final assault—the Apollo program. The United States would experience the ultimate historical first in space exploration: man stepping foot on the Moon and returning safely to Earth.

THE APOLLO PROGRAM

To understand the program as a whole and how the purpose of Apollo changed as it progressed, each of the project's 11 missions is described below. A basic knowledge of each mission is important because, as will become evident in subsequent chapters, the rhetoric used in mediated representations of Apollo oftentimes reflected the mission's objective. Overall, the Apollo program's initial goal was to place humans on the Moon and return them to Earth before 1970.

APOLLO 1

In early 1967, NASA was ready to test the Apollo equipment to inch toward meeting its goal by the end of the decade. On January 27, astronauts Virgil "Gus" Grissom, Edward White, and Roger Chaffee donned their spacesuits for a dress rehearsal that would simulate the first few hours of a mission. Simulating space, the capsule was filled with pure oxygen, which proved to be deadly when combined with a high air pressure of 14.7 psi inside the cabin. Ground instruments alerted engineers that there had been a dangerous rise in oxygen flow, and four seconds later, one of the three astronauts said, "Fire. I smell fire." Two seconds later, another screamed, "Fire in the cockpit!" The spacecraft hatches had been sealed, and the astronauts were overcome with smoke before they could release the emergency hatch, which under normal circumstances would have taken 90 seconds to complete the sequence to release. It took crews outside of the capsule five and a half minutes to open the hatch. While the three men suffered major burns from the fire, it had been the inhalation of toxic gases that had caused their deaths. Needless to say, NASA ordered major design changes and completely revamped rescue response systems in hopes of preventing another tragedy from happening on the launch pad or in space. Apollo 1 was initially designated AS-204 but was renamed Apollo 1 after the fire to honor the three astronauts who had lost their lives.[5]

No missions were designated Apollo 2 or 3. Apollos 4, 5, and 6 were unmanned test flights. Apollo 4, launched November 9, 1967, tested the 3-stage Saturn V rocket. Launched January 22, 1968, Apollo 5 tested the lunar module in space, and Apollo 6 tested the compatibility of the rocket and the spacecraft. It launched April 4, 1968.

APOLLO 7

With astronauts Wally Schirra, Donn Eisele, and Walt Cunningham, Apollo 7 was the first successful manned Apollo test flight that ran from October 11 to 22, 1968. Its purpose was to test the basic Apollo vehicle and its Service Propulsion System to ensure that a trip in and out of lunar orbit was possible for future missions. The mission, later called "a confidence-builder" for the Apollo program, confirmed that the booster, the module, and all the hardware and operations worked perfectly. As such, televised representations of the precision and perfection of NASA's know-how and command of technology illustrated that the United States could reach the Moon before the Soviets, and before the decade would end. The cost of the mission totaled $145 million.[6]

APOLLO 8

Taking place from December 21 to 27, 1968, at a cost of $310 million, the original objective of Apollo 8 was to test the lunar lander in Earth orbit. However, because of the Soviets' threat to become the first to orbit the Moon after its successful unmanned Zond V mission, NASA decided that the Apollo 8 mission would fly to the Moon with no lunar lander attached.[7] Apollo 8 would also mark the first Apollo launch by the massive Saturn V rocket that boosted men to the Moon. In addition to the famous "Earthrise" photograph—humankind's first view of the Earth from another world—Apollo 8 was also the mission that would feature astronauts Frank Borman, Jim Lovell, and Bill Anders in lunar orbit on Christmas Eve while they read verses from the Bible's Book of Genesis, televised to worldwide audiences.

APOLLO 9

The objective of Apollo 9, which flew from March 3 to 13, 1969, was to test the lunar module that would eventually land on the Moon by practicing its performance in Earth orbit, including exercises in undocking and redocking with the command module. Another objective was for astronauts Jim McDivitt, Dave Scott, and Rusty Schweickart to try out a new spacesuit equipped with its own life support system as opposed to the standard "umbilical connection."[8] Because the goal of landing a man on the Moon and returning him safely to Earth depended on a successful orbital rendezvous and docking of Apollo modules, Apollo 9 was paramount at a cost of $340 million.

APOLLO 10

Lasting from May 18 to 26, 1969, Apollo 10 was called "operation dress rehearsal" (CBS, 5/16/69) by Cronkite for the Moon landing scheduled in mid-July that same year. The three Apollo 10 astronauts were Tom Stafford, John Young, and Gene Cernan, and the cost of this mission was $350 million. Other than landing on the Moon, the mission would be exactly as Apollo 11 would be, with precision practices of the Apollo system control and tracking networks, and for surveying the Apollo 11 landing site at the Sea of Tranquility.[9] There was another important first: this mission would enable live color television to be broadcast from space.

APOLLO 11

The Apollo 11 mission, which would fulfill President Kennedy's promise to land an American man on the Moon and return him safely to Earth, took place from July 16 to 24, 1969. This was the primary objective, of course, but a lesser-known goal was landing in unfamiliar territory. It was not known at the time if the lunar lander would sink on the surface of the Moon when it landed. Other firsts included testing the space suits equipped with oxygen and using cameras and devices to drill and collect rock samples on the Moon, to name a few. The "Eagle" touched down at 4:17 p.m. EST on July 20,

1969. Neil Armstrong first made his "one small step" on the Moon six hours later, followed by Buzz Aldrin, and the two spent another 21 hours taking pictures, collecting rock samples, and setting up scientific experiments. As symbolic gestures, the astronauts also planted the American flag and left a plaque signed by the three astronauts and President Richard Nixon that read, "Here men from the planet Earth first set foot upon the Moon. July 1969 A.D. We came in peace for all mankind."[10] This mission would also be the first whereby the two astronauts would launch from the Moon to meet third astronaut Mike Collins in the command ship so that the three could return home safely to Earth, as Kennedy had promised eight years prior. The mission cost $355 million.

APOLLO 12

The goal of Apollo 12 astronauts Pete Conrad, Dick Gordon, and Al Bean was to again land man on the Moon and return him safely home, but this time it was an exercise in the precise targeting of a landing site. Unlike Apollo 11, Apollo 12's LEM landing would be automatic with few manual corrections during landing. Additional goals were to bring back pieces of the Surveyor 3 robot that had landed on the Moon in 1967 for scientific analysis, and to obtain more lunar samples from the Moon's "Ocean of Storms" area.[11] The mission flew from November 14 to 24, 1969, and topped the Apollo 11 mission in cost at $375 million.

APOLLO 13

The initial goal of Apollo 13 was to land in the Fra Mauro region of the Moon, where the astronauts would perform new scientific experiments. However, there was an explosion in the service engine's oxygen tank, and the mission was aborted. After the crew's now-famous announcement, "Houston, we've had a problem," the objective instead became to return astronauts Jim Lovell, Jack Swigert, and Fred Haise safely back to Earth.[12] The mission lasted from April 11 to 17, 1970, at a cost of $375 million even though it was aborted.

APOLLO 14

Lasting from January 31 through February 9, 1971, Apollo 14 accomplished what Apollo 13 had failed to do: land in the Fra Mauro region of the Moon. Scientific packages, including seismic studies and a new pull-cart to carry equipment, were also be new to the mission. While Stu Roosa circled above, Al Shepard and Ed Mitchell were scheduled to collect rocks from the 1,000-foot-wide Cone Crater, but this activity was canceled because the astronauts had difficulty navigating themselves on the Moon. The crew became the first Americans to conduct materials processing experiments in space.[13] The total cost of Apollo 14 was $400 million.

APOLLO 15

The Apollo 15 mission is remembered as the "moon buggy" mission because it was the first to include the lunar rover, a car-like vehicle that enabled the astronauts to cover more lunar territory and therefore conduct more scientific experiments than any other mission. Such experiments produced a significant payoff for the Apollo program. Astronauts Dave Scott, Al Worden, and Jim Irwin brought back a sample of the ancient lunar crust dubbed the "Genesis Rock," one of the oldest rocks known to humankind, and a sub-satellite to measure lunar particles was installed. On August 5, Worden performed the first deep space spacewalk on the return trip. The mission, the longest yet and the most expensive to date at $445 million, took place from July 26 through August 7, 1971.

APOLLO 16

The goal of Apollo 16 was to explore the Descartes highland region of the Moon—thought to provide clues of an extensive volcanism[14]—and to conduct performance tests on the lunar rover. With astronauts John Young, Ken Mattingly, and Charlie Duke, the mission lasted from April 16 to 27, 1972, and cost $445 million.

APOLLO 17

The final Apollo mission, Apollo 17, was also the first mission to place a professional scientist, geologist Harrison "Jack" Schmitt, on the lunar surface with Gene Cernan, while Ron Evans circled in the command module. The objective of the mission was to collect more rock samples, this time from the Taurus-Littrow valley. The astronauts commemorated Apollo's ending by leaving a plaque behind that read, "Here Man completed his first exploration of the Moon, December 1972, A.D. May the spirit of peace in which we came be reflected in the lives of all mankind."[15] With a launch date of December 7 and a return date of December 19, 1972[16] it was the costliest of all the Apollo missions at $450 million.

With a basic knowledge of the Apollo missions, a discussion of how science and technology stories are communicated to the public will be beneficial.

SCIENCE, TECHNOLOGY, AND COMMUNICATION

The Public and Its Science and Technology

The process of communicating science and technology does not occur so much through the works of individual scientists and engineers, but rather as a process that embodies the whole social strata from which a science or technology emerges. As MacKenzie states in *Inventing Accuracy: A Historical Sociology of Nuclear Missile Guidance*, a technology cannot be examined outside of the social structures from which it is created because the discourse is not limited to individuals and engineers.[17] Institutions, organizations, project managers, supporters and non-supporters of a technology, and the media play an equally vital role in how technology is communicated. MacKenzie, for example, describes the invention of accuracy in nuclear missile guidance as a historical product and a social creation, and argues that technological determinism (technology determines society) is a faulty way of examining technology because no technology creates itself or is self-sustaining. In addition, funding sources decide how that technology is refined over time.[18] Even those within an organization disagree on the meaning of a technology. For instance, within the U.S. military, the Army believed the missiles to be a kind of tactical weapon; the Air Force viewed them as counterforce, and the Navy to assure destruction. It then became necessary to convince the public by "social engineering" that the manufacture of the weapon was a positive thing. Through these contests of meaning, the technology was refined in tandem with commercial interests, thereby shifting the meaning yet again. While Americans perceived the atomic bomb to be sufficient to annihilate the world, what got communicated to the public was not the process of creating ballistic missiles, but that the United States was very accurate in producing weapons.[19]

Additionally, Celeste Condit has described how rhetoric is used to communicate science to the public. In *The Meanings of the Gene*, for example, Condit unveiled how the public came to understand genetic research.[20] Taking a rhetorical formation approach, she examined discourses from newspapers, magazines, television, and legislation surrounding public debates concerning genetic research. Just as nuclear-guided missiles fueled and were fueled by the Cold War, so were genetics. Public understanding about the gene was communicated when it became aligned with discourse about the atom. The gene was referred to in the popular press as "the atom of genetics."[21]

Although the atomic bomb created doom and gloom in the minds of the public, the atom itself was thought to be the saving technology against the threat of the Soviet Union. Genetic research was also thought to be the technology that would save the United States from mutations caused by a nuclear attack. According to Condit, the atom as a metaphor greatly increased public understanding of the gene and turned negative perceptions of genetic research into positive ones.[22]

Both authors, MacKenzie and Condit, in their discussions of how technology and science are communicated to the public, assert that there has been doubt among the public in matters as important as who controls life. The public believes that scientists should not have control over such crucial matters; scientists believe that the public is emotional and not well equipped with the knowledge to make rational decisions. In both cases—nuclear missile guidance and genetic research—the debate is solved through market imperatives. With nuclear missiles, commercial uses have caused technologies to become refined and have guided the public to see them as "modern" and therefore necessary. With genetics, the discourse has shifted to human heredity with companies such as 23andMe, for example, that use genetic knowledge to understand and improve one's health, while other organizations enhance global health by producing food to alleviate world hunger and protect the environment. Therefore, knowledge is now "in a thousand different sites,"[23] including commercial, public, and private places, and in many cases these organizations reproduce such information and determine how it gets communicated.

What most people know about science and technology, according to science sociologist Dorothy Nelkin, is what they see in traditional mass media. Today, according to Pew Research Center, 86 percent of U.S. adults turn to a smartphone, computer, or tablet for news. Half receive news from social media and 33 percent from podcasts.[24] It is not through experiences or what they have learned in school that impacts their judgments about science. Because many scientific news stories are laced with complex facts, the general public is thought to need journalistic filters to understand onerous information. But these filters can be problematic. Most journalists are not educated in scientific fields. Thus, they often depend on scientists for "facts" about a particular finding. As Dale Christiansen stated in *Global Warming: A 200 Year History*, scientists, like historians, often bring their own judgments to interpretations.[25] An example of this is the global warming debate. Each side of the issue can "prove" its case with scientific data. Because journalists seek to be objective, they should report both sides of an issue. Oftentimes the controversy becomes the news, rather than the scientific process itself. The danger in this, according to Nelkin, is that the processes of scientific research are omitted, and what the public is left with is confusion and/or no answers at all.

As indicated earlier, another way in which the media have communicated science is through metaphors. Because science is often presented as arcane and is filled with facts beyond the comprehension of most individuals, the media often depend on metaphors to explain the unfamiliar in ways that are familiar to audiences, especially metaphors that reinforce American values. Two such popular examples are frontierism and exploration, which were used repeatedly in reporting the space program because we must "be the first to conquer new territories"; we are "natural explorers"; and "it's the American way." In scientific stories, metaphors also embody the language of war and competition. For example, headlines read, "War on Cancer" and "A Race for the Cure." This can be traced back to 1921 when Edwin Slosson created the first syndicated science column. He thought that science articles should be popularized to appeal to a wide range of readers. Articles consistently employed words with "-est": fastest, hottest, coldest, biggest, and smallest, for example. Today, journalists still report science this way—as competition between nations and between scientists. The problem with reporting science news as a form of sport or a race is that the public will think science is for its own good and thus not argue with findings if it is framed in a way that supports American values. In addition, the press today increasingly use advertising slogans to communicate science and technology, such as "we have the cure for technophobia." Science reporting becomes a source of entertainment instead of information as it is often delivered as promotional instead of critical.[26]

Public relations are even more important in promoting science and technology because of the complexity required to explain esoteric terms and reactions involved in scientific and technological research. No longer do scientists write solely for scholarly publishers. Increasingly they are required to write for the public. "They see gaining national visibility through the mass media as crucial to securing the financial support required to run major research facilities and to assuring favorable public policies toward science and technology," according to Nelkin. [27] While the use of public relations in the United States can be traced back to America's inception,[28] public relations efforts to promote science have greatly increased in recent years. Federal research funding grew after WWII, although scientists did not see a need for relaying scientific information to the public. But after the successful launch of Sputnik in 1957, the United States experienced a resurgence in the importance of relaying information to the public to appease fears about the Soviet Union.[29] NASA devised a very sophisticated public relations office "to attract media attention and win popular support for its costly program."[30] It was not until the tragic Challenger accident of 1986 that the press began to challenge safety and administrative issues and stopped accepting at face value what NASA had spoon-fed them. This, however, goes against the purpose and the rhetoric of the 1958 Space Act, discussed in further detail below, which clearly stated that NASA, by law, must disseminate information pertaining to its activities and results thereof to the public. It also is in direct opposition to what Public Affairs Officers, such as Hugh Harris, who worked at NASA before and during the Apollo years have stated: NASA's mission was not to sell or persuade the public of its importance, but rather to provide information to the public. As Harris stated,

> I don't think NASA felt it had to persuade the public to be supporters of the Apollo program. It really was in the role of providing information to a public that was hungry for information on the subject. We really felt that our job was to provide information and allow the public and leaders of various sorts to make up their own minds.[31]

But how much did NASA have to convince the public to support its goals? When Kennedy made his famous 1961 speech promising to place a man on the Moon before the decade ended, public support for the Moon program was mostly positive, as citizens hoped that the United States would rush ahead of the USSR in the Space Race after the Russians' success with Sputnik in 1957. Americans supported the newly formed space agency and its Moon mission, though not overwhelmingly, and public approval was the most positive for Apollo 11. Support fell drastically after the climax of Armstrong and Aldrin walking on the Moon, although there was a minor spike in interest for the dramatic Apollo 13, and later for Apollo 15 and its Moon buggy, and finally for Apollo 17 because it was the last Moon mission.

COMMUNICATING SPACE

In the beginning of the Space Race, the risks of space travel were downplayed, and the benefits of being first were magnified. Perhaps the creators and managers of all large government-funded scientific projects have cautioned against adversity. Even J. Robert Oppenheimer, who directed the building of the first atomic bomb with the Manhattan Project, was warned by a colleague not to publicly discuss the risks of atomic warfare. Though Oppenheimer later offered moral arguments against nuclear warfare, he was a popular hero after World War II and replaced Albert Einstein as "the public image of genius."[32] Like the Manhattan Project, NASA funding was backed by Congress based on the argument of America's survival and well-being. Similarly, in 1960, German-born rocket scientist Wernher von Braun warned Congress against Soviet successes in space: "I fear it is later than we think, and our position in the world is gravely endangered."[33]

According to Mark Byrnes, von Braun's argument is an example of nationalism, an image that portrays space exploration as providing benefits to America and appealing primarily to the American "love of country."[34] While NASA's public relations campaign will be discussed in chapter 3, it should be noted here that NASA used several strategies to promote its programs. In *Politics and Space: Image Making by NASA*, Byrnes examines NASA's history in three phases: the Mercury era (1958–1963), the Apollo era (1964–1972),[35] and the Shuttle era (1973–1990). Within these three segments, Byrnes identifies the dominant images and rhetoric used by NASA and argues that the space agency purposely chose these images and rhetoric through its public relations apparatus to maintain public support. Nationalism, most prevalent in NASA's early years during the Mercury program, was NASA's answer to the Soviet's Sputnik, and depended on themes of national pride, national prestige, national strength, and international peace and cooperation.[36] NASA maintained the image of nationalism by affirming a Soviet threat and the imminent danger of the United States falling behind technologically. It also emphasized the fact that the United States must be proactive in its space quests to maintain national security. While these themes were evident in the Apollo project, according to Byrnes, romanticism was most prevalent during the Apollo years, with images of exploration, heroism, and emotional rewards, focusing on challenge, inspiration, adventure, imagination, spirituality, and satisfying curiosity.[37] Romanticism "invited the public to participate vicariously in NASA's exciting and heroic activities," emphasizing "the spectacular nature of NASA's endeavors" to offer "the public a way to escape temporarily the humdrum routine of daily life."[38] Nationalism was also a dominant image during the Apollo era, especially up until the Moon landing in 1969, when America had clearly won the space race.

Once Americans became accustomed to the successful Moon landings, NASA then shifted to an image of pragmatism, and its officials turned to convincing the public of space exploration payoffs.[39] Operating under the assumption that space exploration generates useful knowledge that improves the daily life of citizens, NASA's use of pragmatism encouraged the public "to rely on rational calculation rather than emotion in assessing the value of the space program."[40] This is consistent with Nelkin's research that shows how organizations accelerate campaigns to persuade the public that funding is worthwhile for scientific and technological research. However, it did not work for Apollo. The last three Apollo missions were canceled amid strong public opposition to additional Moon excursions by 1970. Interestingly, this was also a midterm election year that influenced Richard M. Nixon, who, at one time, claimed that the Apollo 11 mission was "the next best thing to creation." By the time the Shuttle era arrived, NASA almost completely relied on pragmatism in the 1970s and 1980s, which were lean budget years for the space agency. After the Challenger tragedy in 1986, nationalism became the dominant image from 1986 to 1990[41] in hopes to regain the public support that waned after the accident.

In addition to being historical firsts that belonged to America, the Moon shots were the "right stuff" for good television. Simultaneously, Apollo offered drama, adventure, tragedy, heroes, suspense, and education—all part of television's enabling viewers to witness history in the making. Above all, space exploration, as acted out on television, provided an outlet for the American myths of hero and frontier during a time when its young men were dying in the jungles of Vietnam, and its sons and daughters were fighting for equality on several fronts on the streets of cities across the United States. While the tidy men of NASA's engineering force did their share of fighting the Cold War in the safety of what NASA's brass called the "trench" of Mission Control, the Apollo project underscored to the American public, and to the world, that the United States government was in control; that through the technology of its mighty rockets and clean-cut astronauts and engineers, nothing was impossible, especially winning the Cold War and beating the Russians at their own game.

Indeed, two of the most prominent narratives of the Apollo mission centered upon hero and frontier. Both of these narratives fit into a mythic profile. As John Hartley explains in *Understanding News*, the function of a myth is:

> to allow a society to use factual or fictional characters and events to make sense of its environment, both physical and social. . . . They endow the world with conceptual values that originate in their language. News is a mythmaker.[42]

Similarly, Richard Campbell showed in his study of *60 Minutes* that both myths and news stories help us deal with and make sense of everyday life.[43] The same can be said of *CBS Evening News*, CNN, THC, and PBS, all mythmakers and mythkeepers reliant especially on the myth of the frontier and the hero. Even the project's name, Apollo, suggested by NASA's Director of Space Flight Development Abe Silverstein, was taken from Greek mythology as Apollo was the god of the sun who, in his horse-drawn chariot, pulled the sun across the sky each day.[44]

Perhaps the most celebrated myth in American history has been Frederick Jackson Turner's Frontier Thesis. Turner argues that America has been chiefly influenced by its contact with the wilderness and "that our national customs and character, indeed our sources of success as a people, were largely a product of our frontier experience."[45] The frontiersman is the crucial figure in the Frontier Thesis,[46] and this "national hero [with his] mythic character was capable of solving virtually any problem facing Americans, at any time."[47] Still others, such as James Kauffman, describe factors specific to the U.S. manned space program that originated from Kennedy's administration, aptly christened "New Frontier." The "New Frontier" evoked "a rugged independent pioneer who must travel through a hostile, unknown environment and overcome a malevolent antagonist in his quest to reach a specific, geographic location capable of being conquered and dominated."[48] Characterizing the astronaut as a frontiersman was evident in American media since the inception of the space program and continues, for it was the astronaut's walk on the New Frontier of the Moon that would save humanity from the wicked communist empire.

Many believe that the frontier myth is alive and well today. In her analysis of western films, Janice Rushing characterized the 1980s as "the era of the urban cowboy," "for favoring western clothing and music, emulating frontier customs, and electing a 'cowboy President' by a landslide."[49] As Ronald Carpenter argues, the frontier myth undeniably thrives in American culture, especially in advertising and the branding of products. Automobile advertising features names like Ford "Bronco," Dodge "Dakota," and Chevrolet "Blazer," and depicts these rugged oversized vehicles conquering rough terrain with the most majestic and open landscapes in America as backdrops.[50] Ironically, only 6 percent of the owners of these vehicles are ranchers.[51]

Just as John Kennedy asserted more than 40 years ago, no longer is the frontier confined to the Earth's land and sea. Citing Kennedy's rhetoric of his "'New Frontier' call to explore the stars, conquer the deserts, and eradicate disease"[52] as the United States entered space, Carpenter provides example after example of Hollywood portraying the New Frontier of space modeled on the frontier of the Old West. For example, the movie *Outland* is "essentially a remake of *High Noon*, replacing Gary Cooper with Sean Connery" as the marshal, and the outpost became the Moon.[53] In *The Right Stuff*, Chuck Yeager "frequents Pancho's Fly Inn, 'the saloon in any classic western,' and rides a horse to the X-1 in which he will scout space's perimeter."[54] These examples multiply following the Apollo era in space westerns such as *Star Wars*, where characters are seen with laser guns holstered below the hip and where characters visit space saloons borrowed from classic cowboy movies.[55] Films such as *Space Cowboys* (2000), though a satire of the traditional young heroes, featured an older "cowboy" Clint Eastwood, a western genre star. Similarly, the space western film *Armageddon* (1998) featured Bruce Willis who starred a decade earlier in *Die Hard*, where terrorist antagonist Hans Gruber referred to Willis's character, John McClane, as "Mr. Cowboy," followed by McClane's famous "Yippee-Ki-Yay . . . ," a line taken from the Bing Crosby song *I'm an Old Cowhand*, and later popularized by cowboy actor Roy Rogers on his 1950s *Roy Rogers Show*.

In addition to appearances in cultural products such as advertising and film, frontier rhetoric is employed by even the largest of American cultural institutions, the federal government. Presidents have long appealed to the American public's sense of frontierism, and such symbolism can certainly be found in the rhetoric of President George W. Bush, a rancher from Texas, both in his speeches to the nation pertaining to space exploration and in his 2004 presidential campaign slogans. For example, billboard ads across the United States featured a slogan beneath nothing more than an open prairie scene: "Boots or Flip-Flops? Vote Bush-Cheney."[56] Bush had also invoked frontierism in numerous speeches, including his promise to the nation a trip back to the Moon and then onto Mars. Here, Bush's rhetoric was laced with frontier metaphors—"Mankind is drawn to the heavens for the same reason we were once drawn to unknown lands and across the open sea;" and "We will build new ships to carry men forward."[57] The phrase "unknown lands" recalls images of covered wagons making the journey across promising territories, while "new ships" and "open seas" conjure up images of Christopher Columbus sailing the ocean toward new land. Bush also mentioned two familiar frontiersmen:

> Two centuries ago, Meriwether Lewis and William Clark left St. Louis to explore the new lands acquired in the Louisiana Purchase. They made that journey in the spirit of discovery, to learn the potential of vast new territory, and to chart a way for others to follow. America has ventured forth into space for the same reasons. We have undertaken space travel because the desire to explore and understand is part of our character.[58]

To capitalize on the virtues of "one mission, one voyage, one landing at a time," Bush detailed medical and communication technologies that have resulted from the space program. Now, instead of a westward frontier, Bush looked upward "to explore space because doing so improves our lives and lifts our national spirit."[59] The myth of the frontier, then, serves to draw Americans together for a common good—to move the country forward. Americans, of course, can approve of space exploration when it is presented in terms of the symbols they hold dear, just as they did when Kennedy spoke of the nation's space effort at Rice University in 1962, "This country was conquered by those who moved forward—and so will space [be conquered by those who move forward]."[60] With more than 40 years separating their speeches, Kennedy and Bush both realized the importance of the frontier as symbolic to America's "land that I love."

MYTHS

Ernest Bormann suggests that myths "serve to sustain the members' sense of community, to impel them strongly to action (which raises the question of motivation), and to provide them with a social reality filled with heroes, villains, emotions, and attitudes."[61] Supplementing the myths of space as the new frontier and the astronaut as the frontiersman is the myth of the hero, another image constructed by NASA and the media to present the space race to the public. As space historian Roger Launius has argued, it was only after NASA unveiled the first Americans to fly in space on April 9, 1959, that the astronauts became "heroes" in the public mind. Before that time, "they were a crew-cut, military-minded, mad-monk, thrill-seeking, hard-drinking, woman-chasing, flying-fool gang of daredevils."[62] While the focus on astronauts being military test pilots had not been NASA's intention, President Eisenhower asked that the astronauts be chosen from the pool of U.S. armed forces test pilots. Historian Margaret Weitekamp explains that Eisenhower's preference "gave NASA a way to tap into a self-selected group of men who were trained to think like engineers and had already volunteered to risk their lives for their country."[63] The heroic deed of volunteering for the United States seemed to be an underlying theme in Cold War mentality, mirroring Kennedy's "Ask not what your country can do for you, but what you can do for your country" trope in his inaugural speech. As we will see, the space program and its chosen participants were portrayed, especially in the early years, as those with a steadfast love for God, country, and family—true American heroes in the minds of the public.

Implicit was the conception of the hero as an ideal citizen—one who followed the rules and did not question authority. As Weitekamp explained,

> Through that [military test] flight experience, the jet pilots also mastered valuable skills that NASA wanted its astronauts to possess. Test pilots flew high-performance aircraft to detect problems, diagnose the causes, and clearly communicate the analyses to the engineers and mechanics. In addition, military pilots understood discipline, rank, and order.[64]

Above all, they would be able to take orders, a quality embedded in the work and family ethic of Middle America in the 1950s and early 1960s but would later be challenged by America's youth in the mid- to late-1960s.

Given these myths of frontierism and heroism, a look at media scholar James Carey's work will guide an understanding of how these narratives were communicated on television. Carey's ritual model of communication might also explain how such American myths have lasted for decades in the media, beginning with Apollo's first successful flight to the Moon in 1968. As Ronald Carpenter lamented, "History's presentation of the past mold myths which are bases of action for the future."[65]

RITUAL MODEL OF COMMUNICATION

In *Communication as Culture*, James Carey writes that communication can be divided into two conceptions, the transmission and the ritual view of communication. According to Carey, the transmission view, more prevalent in American culture than the ritual view, follows traditional models of communication whereby information is communicated (transmitted) from a sender to a receiver. It regards communication as a "process of transmitting messages at a distance for the purpose of control" to persuade, change attitudes, or modify behavior in society.[66] Thus, the transmission view concentrates on the message.

Communication also serves as a method by which a society's ideals are created and reinforced. This alternative mode of communication, the ritual model, is less concerned with "imparting information" and is more concerned with "the representation of shared beliefs."[67] This "more literary"[68] approach regards communication as "not directed toward the extension of messages in space," but rather "the maintenance of society in time."[69] Bush's speech about frontiers can be construed as transmission because he argues that we should go back to the Moon for technological benefits derived from exploring it. Taken as a ritual message, however, the meaning is that we should go back to the Moon because it is "the American thing to do." The message is one of common values: "Americans are natural explorers," or "We should ban together as Americans." Carey uses words such as "sharing," "participation," and "association,"[70] to describe the ritual view of communication because beliefs held by society are shared through its symbols, such as the American flag, for example. Indeed, as Bush made his lunar return pitch, an American flag served as the backdrop behind his podium in prominent view. Institutions that represent art, science, journalism, religion, and mythology, for example, generate symbols that are shared in the community, drawing its members together in "fellowship and commonality."[71] That function, says Carey, serves a primal need as "communication is a symbolic process whereby reality is produced, maintained, repaired and transformed."[72]

To illustrate, the following two examples incorporate Carey's framework of the ritual view of communication and are relevant to this discussion because they examine contemporary media coverage of events that were prominent in the late 1960s and early 1970s, like Apollo. In "'Come Together, Over Me': Generational Memory and Social Drama in the Death of John Lennon," Fred Fogo argues that the United States in the 1990s was attempting to mend many of the cultural and political conflicts of the 1960s. Drawing on Carey's ritual view of communication, Fogo examines these processes by looking at the symbolic meaning attached to John Lennon after his death, explaining how the sixties generation made sense of everyday life through symbols attached to John Lennon and his music.[73] Similarly, in "Mediated Vietnam: The Politics of Postwar Representations," Harry Haines examines media representations of Vietnam War combat veterans. Applying the ritual view of communication, Haines seeks to explain what it means to be a Vietnam War hero and how televised representations such as *Rambo*, *Magnum, P.I.*, and documentaries of the war have contributed to these meanings. Haines suggests that these representations are symbolic of an ideological struggle between the soldier as "hero" and the soldier as "witness." These representations, as cultural forms, ritualize the meaning of the Vietnam War, as do more contemporary sites of discourse such as the Vietnam Veterans Memorial in Washington, D.C.[74]

While the transmission view addresses the message, the ritual view, on the other hand, tends to follow a cultural studies tradition in that it is more concerned with the overall ideology of a society and how that ideological power is exercised in a democracy. But one view must not discount or exclude the other. Rather, to correctly comprehend communication as a process, the ritual view of communication must be considered. Following this line of thought, to fully understand the reporting of the Apollo program, one must not analyze only the messages broadcast on television, but also the meaning that these messages impart through metaphors and the ideology behind the reporting of these stories.

The ritual model is compatible with Pamela Shoemaker and Stephen Reese's notion that American media studies should be more concerned with *how* the content produced in mediated messages is influenced by factors such as ideology than with the grand generalization of audience perceptions of those messages. The message does not lie solely within the information itself but also within the ritual aspects of its communication. The ideology or symbolism that shapes the message and affirms its meaning is just as important. As Shoemaker and Reese explain, "Not only is news about the powerful, but it structures stories so that events are interpreted from the perspective of powerful interests."[75] Using their example, what does it mean that more optimistic Pentagon reports of frontline Vietnam War action were selected over reporter David Halberstam's pessimistic reports based on actual on-the-scene experiences?[76] The problem was that the conventional reporting of the Vietnam War, aimed at creating a sphere of consensus, in turn purged any discourse of the moral and political implications of the war.[77]

COLLECTIVE MEMORY

Collectively sharing memories is a way to "characterize the past in ways that merge the past, the present, and the future into a single, ongoing tale."[78] While collective memory narratives often focus on tragic rather than triumphant events in history—events that we might want to forget because they were painful experiences—they are also remembered because they teach a valuable lesson. These events include the Holocaust, the Kennedy assassination, the Vietnam War, the Watergate scandal, the Challenger explosion, and 9/11. Because these events are commemorated by the media on anniversaries of their occurrence, they are kept alive in popular memory. Triumphant events also unite society, define culture, and ultimately reinforce symbols widely shared in American culture. This is important because the way in which a society comes to think of a particular event subsequently influences policy. For example, as Michael Schudson illustrated in *Watergate in American Memory: How We Remember, Forget and Reconstruct the Past*, Watergate, even today, influences law and defines political scandals. In the case of Apollo, American ideals of the frontier, of being "first" or winning, and of pride in technological innovation influence policy.

"Collective memory" thus can function as a theoretical context for studying representations of Apollo. Maurice Halbwachs, known as the father of the collective memory concept, developed it as a philosophical basis for understanding how groups of people remembered. At a time when psychologists were interested in individual memory, Halbwachs believed that "individual consciousness is not only the way in which personal identity is constituted; contacts with other people and with other things can supplement it."[79] For him, memory is a socially constructed *process* wherein the past is evoked by institutional consciousness as well as through historiography and biography.[80] While individual memories are limited in time and space, collective memories encompass a wider temporal and spatial range. Halbwachs believed that collective memory "retains from the past only what still lives or is capable of living in the consciousness of the groups keeping the memory alive,"[81] and warned that it should not be confused with history.

Discussions of collective memory have been expanded. No longer is it limited to history and psychology disciplines, but rather spans communications, law, geography, and folklore, all of which use collective memory as a theoretical base in research.[82] Defining collective memory as "the full sweep of historical consciousness, understanding, and expression that a culture has to offer,"[83] Gary Edgerton has indicated that collective memory is "the site of mediation where professional history must ultimately share space with popular history."[84] Referring to how history is presented to the masses via the media, Edgerton recognizes this area of scholarship as important because people obtain historical information from television and base their understanding of history on such televised representations.[85] Today, social media acts as the conduit for historical remembrances, sans gatekeepers, by which people from around the world share their lived experiences of events.

In sharing memories, Schudson wrote that, ". . . a nation 'remembers' institutionally as well as individually. . . . Remembrance is collective—institutional and cultural, located in language and legislation."[86] He defines collective or social memory as "the ways in which group, institutional, and cultural recollections of the past shape people's actions in the present."[87] In addition, Schudson identified six characteristics of collective or social memory. First, memory is not limited to individual minds; it is sometimes located in institutions. Second, collective memory is located in and created by monuments and markers such as holidays or statutes. Third, collective memory can be located within certain generations or groups. Fourth, collective memories are social and cultural because they operate through cultural constructions of language and social cues. Fifth, collective remembering is almost always social and interactive: social situations and cultural artifacts prompt memory. And sixth, collective memory is variable; different cultures have unique ways of remembering and forgetting.[88] The narrative in the research at hand will focus on each of these six in the pages to come. Similarly, Barbie Zelizer defines collective memory as "memories of a shared past by those who experienced it and whose conscious and strategic efforts have kept it alive."[89] She argues that the media have had a distinctive role in shaping America's collective memory of President Kennedy and in the process have promoted themselves as cultural authorities. This notion of cultural authority is paramount in a topic such as the one at hand because the media as cultural authorities determine what is said about an event. In sum, media are institutions through which memory is created and maintained.

However, while collective memory is a useful theoretical tool, its focus on the audience is a bit problematic for the research at hand. Theory is two-sided. One side examines the memory of many individuals who together create a "collective memory." The other side sees institutions in society as creating memory that fills the gaps or readjusts individual memories. Because the aim of this research is to determine how the media, specifically television, presented Apollo, there should be less concern with audience perception and memory. In addition, because rhetoric and meanings of technologies are always contested and therefore always evolving, there can never be one meaning or one coherent memory of a specific technology.[90] Apollo, for example, was a means by which to beat the Russians and to show the superiority of a free capitalist world to some, a waste of money to others, and the epitome of progress to still others. As postmodernists point out, there are many different sites of discourse, or "metanarratives,"[91] as Jean-Françoise Lyotard termed them, and reconstructing them all would take us too far afield. Thus, rather than analyzing individual viewpoints, a critical analysis of any cultural medium should instead "define and describe the inventory that makes possible the multiple meanings extracted by audiences, creators, and network makers,"[92] according to Horace Newcomb and Paul Hirsch.

REMEMBERING APOLLO THROUGH RITUAL COMMUNICATION

While there are collective memory stories and oral histories of the lived experiences of those who worked on or witnessed Apollo, these memories may have faded with time or changed as they have seen more recent representations on television, the silver screen, and the Internet. Therefore, rather than focusing on what Apollo meant for the audience, an emphasis will be placed on Carey's model, which seems to be a more precise extension of collective memory because, to draw these individuals together, there must be "collective values and symbols of identification."[93] The national media identify for the public that which enables a common understanding to be reached. In this respect, the mass media "become even more important than face-to-face interaction in the processing of information within the group, in the assigning of status, and in the development of collective ideologies and values."[94]

Useful also is Carey's discussion of a "professional communicator" who translates the symbols the media make available to us. This person, a "broker of symbols," is one who "controls a specific skill in the manipulation of symbols . . . to forge a link between distinct persons or differentiated groups."[95] Professional communicators convert messages into symbols for the masses. They are not only transmitters of information but also of symbols. These communicators are important, according to Carey, because they tie together common communication channels and create shared perspectives through their interpretations. In this regard, Walter Cronkite, a firm believer in the space program, shared his appreciation of Apollo and related the significance of Apollo to his audiences. Additionally, as an individual worker within the CBS organization, Cronkite became tremendously influential in producing the content of Apollo news stories. Therefore, memory as a cultural production akin to Carey's ritual model of communication will be employed. The result is an extension of collective memory theory, not by looking at how individual memory evolves, but rather at how cultural institutions contribute to the creation and maintenance of shared beliefs involving a national event. How the story gets told in turn influences how society comes to think of it.

It is through cultural institutions that memory is created, transformed, and maintained. Some might argue that professional historians are part of a cultural institution that keeps memory alive on any given subject. This is true, but it is also through mass media institutions such as network news, cable news, documentaries, and public television that Apollo is kept alive.

Language that is heard and images that are seen in the mass media keep the Moon shots active in the collective minds of society. The broadcast media have marked anniversary dates of space exploration milestones and, equally important, have rebroadcast footage of these milestones as each of the 24 Apollo astronauts, "American heroes" all, passes away because viewers have shared a part of their lives with them vis-à-vis television. But Apollo is also recreated with every discourse about another Moon mission—whether portrayed as a race or a journey—as it brings to mind the initial thrill created more than 50 years ago.

NOTES

1. Linda Krug, *Presidential Perspectives on Space Exploration: Guiding Metaphors from Eisenhower to Bush* (New York: Praeger, 1991), 25. See also Walter A. McDougall, *. . . the Heavens and the Earth: A Political History of the Space Age* (New York: Basic Books), 128.
2. Pam Parry, *Eisenhower: The Public Relations President* (Lanham, MD: Lexington Books, 2016), 98.
3. Krug, *Presidential Perspectives on Space Exploration*, 30.
4. Roger D. Launius, *Apollo: A Retrospective Analysis, Monographs in Aerospace History* 3 (Washington, DC: National Aeronautics and Space Administration, 1994), 5.
5. "The Apollo 1 Tragedy," National Aeronautics and Space Administration, accessed May 21, 2020, https://nssdc.gsfc.nasa.gov/planetary/lunar/apollo1info.html.
6. *History of Manned Space Flight*, NASA Publication #75-24641 (Washington, DC: National Aeronautics and Space Administration, 1975), 6.
7. Launius, *Apollo*, 27.
8. Ibid., 28.
9. Ibid.
10. Ibid.
11. Ibid., 29.
12. Ibid.
13. Ibid.
14. Michael Collins, *Carrying the Fire: An Astronaut's Journeys* (New York: Farrar, Straus, and Giroux, 1974), 482.
15. Launius, *Apollo*, 30.
16. CBS Evening News' Walter Cronkite reported on August 28, 1972, that Apollo 17 would be the last of the Apollo missions.
17. Donald MacKenzie, *Inventing Accuracy: A Historical Sociology of Nuclear Missile Guidance* (Cambridge, MA: MIT Press, 1990).
18. Ibid.
19. Ibid.

20. Celeste Condit, *The Meaning of the Gene: Public Debates about Human Heredity* (Madison, WI: The University of Wisconsin Press, 1999).

21. Ibid.

22. Ibid.

23. Ibid.

24. Jeffrey Gottfried, "Americans' Social Media Use," Pew Research, January 31, 2024, https://www.pewresearch.org/internet/2024/01/31/americans-social-media-use/.

25. Gale E. Christenson. *Greenhouse: A 200-Year History of Global Warming* (New York: Penguin Books, 2000).

26. Dorothy Nelkin, *Selling Science: How the Press Covers Science and Technology* (New York: W.H. Freeman and Company, 1987), 89-90.

27. Ibid., 133.

28. Dennis L. Wilcox and Glen T. Cameron, *Public Relations: Strategies and Tactics,* 8th ed. (Boston, MA: Pearson/Allyn and Bacon, 2006), 22. For example, The Virginia Company began a campaign in 1620 to persuade immigrants to relocate to Virginia by offering 50 acres of free land. Once American colonies were established, public relations techniques were employed to promote institutions such as Harvard in 1641 and Kings College (now Columbia University) in 1758. For another excellent history of public relations in the United States, see Stuart Ewen's, *PR! A Social History of Spin* (New York: Basic Books, 1996).

29. Nelkin, *Selling Science*, 135.

30. Ibid., 137.

31. Hugh Harris, interview with author, April 2, 2004.

32. Jane Gregory and Steve Miller, *Science in Public: Communication, Culture, and Credibility* (New York: Plenum Trade, 1998), 36.

33. United States Congress, Senate, *Committee on Aeronautical and Space Sciences, Subcommittee on NASA Authorization, NASA Authorization for Fiscal Year 1961, Part 1, 86th Congress, 2nd Session*, March 29, 1960, 241. Accessed May 14, 2024, https://www.nasa.gov/history/SP-4002/p1a.htm.

34. Mark E. Byrnes, *Politics and Space: Image Making by NASA* (Westport, CT: Praeger Publishers, 1994), 7.

35. Some historians mark the beginning of Apollo in 1961 when Kennedy proposed landing on the Moon.

36. Ibid., 8.

37. Ibid., 64.

38. Ibid., 47.

39. Byrnes, *Politics and Space*, 170.

40. Ibid., 99.

41. Ibid., 172.

42. John Hartley, *Understanding News* (London: Metheun Publishing Ltd., 1982), 30.

43. Richard Campbell, *60 Minutes and the News* (Urbana, IL: University of Illinois Press, 1991).

44. Richard L. Lattimer, *All We Did Was Fly to the Moon* (Gainesville, FL: Whispering Eagle Press, 1985).

45. Ronald H. Carpenter, "Frederick Jackson Turner and the Rhetorical Impact of the Frontier Thesis," *Quarterly Journal of Speech* 63 (April 1977): 117.

46. Ibid., 120.

47. Ibid., 125.

48. James Kauffman, "A Successful Failure: NASA's Crisis Communications Regarding Apollo 13," *Public Relations Review* 27 (2001): 442.

49. Janice Hocker Rushing, "Mythic Evolution of 'The New Frontier' in Mass Mediated Rhetoric," *Studies in Mass Communication* 3 (1986): 265–296.

50. Ronald H. Carpenter, "America's Tragic Metaphor: Our Twentieth-Century Combatants as Frontiersmen," *Quarterly Journal of Speech* 9 (February 1990): 14.

51. Ibid.

52. Carpenter, "Frederick Jackson Turner," 10.

53. Ibid.

54. Ibid.

55. Ibid.

56. "Flip-flops" in the ad refers to the criticism from the political right that presidential candidate John Kerry "flip-flopped" on issues in his role as U.S. Senator.

57. "Bush Sets 'New Course' for the Moon and Beyond," *NBC News*, accessed June 19, 2024, https://www.nbcnews.com/id/wbna3950099.

58. "President Bush Announces New Vision for Space Exploration Program," The White House, accessed June 18, 2024, https://georgewbush-whitehouse.archives.gov/news/releases/2004/01/20040114-3.html.

59. Ibid.

60. "Address at Rice University on the Nation's Space Effort," John F. Kennedy Presidential Library, accessed June 12, 2024, https://www.jfklibrary.org/learn/about-jfk/historic-speeches/address-at-rice-university-on-the-nations-space-effort.

61. Ernest G. Bormann, "Fantasy and Rhetorical Vision: The Rhetorical Criticism of Social Reality," *Quarterly Journal of Speech* 58 (1972): 398.

62. Roger D. Launius. "Heroes in a Vacuum: The Apollo Astronaut as Cultural Icon," *Florida Historical Quarterly* 87, no. 2 (2020): Article 6. https://stars.library.ucf.edu/fhq/vol87/iss2/6.

63. Margaret Weitekamp, *Right Stuff, Wrong Sex: America's First Women in Space Program* (Baltimore, MD: The John Hopkins University Press, 2004), 42.

64. Ibid.

65. Carpenter, "Frederick Jackson Turner," 118.

66. James Carey, *Communication as Culture* (New York: Routledge, 1989), 18.

67. Ibid.

68. Horace M. Newcomb and Paul M. Hirsch, "Television as a Cultural Forum: Implications for Research," in Willard D. Rowland, Jr. and Bruce Watkins, eds. *Interpreting Television: Current Research Perspectives* (Beverly Hills, CA: Sage Publications, 1984), 60.

69. Carey, *Communication as Culture*, 1.

70. Ibid., 18.

71. Ibid. 30.

72. Ibid., 23.

73. Fred Fogo, "'Come Together, Over Me': Generational Memory and Social Drama in the Death of John Lennon" (PhD Dissertation, The University of Utah, Salt Lake City, UT, 1990).

74. Harry Haines, "Mediated Vietnam: The Politics of Postwar Representations" (PhD Dissertation, The University of Utah, Salt Lake City, UT, 1987).

75. Pamela Shoemaker and Stephen Reese, *Mediating the Message: Theories of Influences on Mass Media Content*, 2nd ed. (White Plains, NY: Longman Publishers, 1991), 224.

76. Ibid., 245.

77. Ibid.

78. Carolyn L. Kitch, *Pages from the Past: History and Memory in American Magazines* (Chapel Hill, NC: The University of North Carolina Press, 2005), 11.

79. Maurice Halbwachs, *The Collective Memory* (New York: Harper Colophon, 1950), 5.

80. Maurice Halbwachs, *On Collective Memory*, trans. Lewis A. Coser, ed. (Chicago, IL: University of Chicago Press, 1992), 2–3.

81. Ibid., 80.

82. David Middleton and Derek Edwards, *Collective Remembering* (London: Sage Publications, 1990), 3.

83. Gary R. Edgerton, "Television as Historian: A Different Kind of History Altogether," in Gary R. Edgerton and Peter C. Rollins, eds. *Television Histories: Shaping Collective Memory in the Media Age* (Lexington, KY: University of Kentucky Press, 2001), 1.

84. Edgerton, "Television as Historian," 5.

85. Ibid.

86. Michael Schudson, *Watergate in American Memory: What we Remember, Forget, and Reconstruct about the Past* (New York: Basic Books, 1992), xiii.

87. Ibid., 3.

88. Ibid., 51–52.

89. Barbie Zelizer, *Covering the Body: The Kennedy Assassination, the Media, and the Shaping of Collective Memory* (Chicago, IL: The University of Chicago Press, 1992), 3.

90. Condit, *The Meaning of the Gene;* MacKenzie, *Inventing Accuracy.*

91. Jean-Francois Lyotard, *The Postmodern Condition: A Report on Knowledge* (Minneapolis, MN: The University of Minnesota Press, 1984).

92. Newcomb and Hirsch, "Television as a Cultural Forum: Implications for Research," 71.

93. Carey, *Communication as Culture*, 25.

94. Ibid., 26.

95. Ibid., 27.

Chapter 2

Rhetorical Analyses and Televised Metaphors

On the morning of July 16, 1969, the day the United States launched the spacecraft that would land man on the Moon to place his first historical footsteps there, the following televised conversation took place between CBS News correspondent Walter Cronkite and CBS News commentator Eric Sevareid:

> Cronkite: I've noticed in the reporting that those under 16, who've really grown up with space since its first memorable moments—when they were four or so and the space thing was just coming into being—understand it and want more detail in our reporting. They want to know about escape velocity, and they want to know about the lunar trajectory velocity, and those over 30 or so say, "Don't tell me all that, I just don't understand. Tell me when we get there."
>
> Sevareid: Furthermore, this is not a romantic era, not a poetic era. The beauty the young find is in the things themselves. All the imagery and the words will come later, but we really don't have a language to describe this thing. As we sit here today, what are the words you use? I think the language is being altered, many new words and phrases and concepts are being added, and, I think, some language is being eliminated. How do you say, "high as the sky" anymore, or "the sky is the limit"—what does that mean?
>
> Cronkite: Maybe it's that we have been so busy, so many things crowding in on us, we haven't had time for language.
>
> Sevareid: There's always a great cultural lag on these things. It takes a long time for a new language to emerge.[1]

As David Eason has stated, reporting news is a linguistic and cultural act because language "mediates between the event and the possible forms which the story of that event may take within culturally provided categories."[2]

Now that we have five decades of history to reflect upon what language has emerged since the "glory" days of NASA, we can examine what language has been created, what language has been omitted, and what language has remained in stories of space exploration, particularly those repackaged for new audiences. What are the culturally provided categories Apollo provided and continue to evoke today? As noted, the ways in which shared experiences are talked about in culture thereby determine the memory of it.[3] Thus, the narratives used to describe the Apollo Project as an American event shaped the memory of it, and in these renewed representations of space exploration, the language has changed, as will be evident throughout this book.

A rhetorical analysis has enabled an interpretation of televised representations by examining patterns of narratives and metaphors. Television seems well suited for this type of analysis because, as Gary Edgerton has noted, "unlike written discourse, the language of TV is highly stylized, elliptical (rather than linear) in structure, and associational or metaphoric in the ways in which it portrays images and ideas."[4] Additionally, a rhetorical analysis, based in the humanistic tradition, seeks to "unpack and reveal meaning in the symbolic conventions used by communicators" and seeks to "explicate the manner in which a communicator's rhetorical purpose is manifest in various semantic and structural elements of persuasive communication."[5] As Bush and Boller have indicated, rhetorical analyses also enable researchers to uncover the ideologies of communicators, which is paramount. Burke's method of rhetorical analysis posits that communicators, in producing content, purposely select conventions that emphasize certain elements while "revealing their ideological assumptions about the specific human situation with which they are concerned."[6] The ideologies of NASA and television networks have contributed to the ways in which Apollo has been presented.

Still others have identified the importance of using rhetorical analyses in examining science and technology because science and technology are so closely tied to social relations. In analyses of cultural, legal, and scientific rhetoric involving the Human Genome Diversity Project, Hasian and Plec, like Nelkin, Condit, and MacKenzie, indicate that the way in which science is discussed often determines the negotiation of technical and social interests and often affects the perception that a scientific project is feasible and desirable among the public.[7] Because large scientific projects require a positive opinion among the populace to be funded, they must be explained in persuasive terms by communicators.

It is important to examine television as an object that, "through its storytelling functions, unites and examines a culture."[8] Television is also "the ideal facilitator of cultural memory, with its ritualistic, event-style coverage and capacity for endless repetition."[9] Although it is primarily a visual medium, with television news, sound dominates.[10] According to sociologist and media critic Herbert Gans, the verbal sound is what brings the images together in television news.[11] Similarly, Raymond Williams has noted that images make no sense without sound; it is the sound and the image together that make news stories "flow."[12] Despite the fluidity of a news story, however, a drawback of using television to examine particular events is that an audience's oral or visual reception of a message cannot be measured. While some audiences may interpret a message to have only one meaning, that same message may create multiple meanings for which the researcher may not have controlled. Here, I am more interested in the ideology of *why* and the way in which stories are told, rather than in audience reception. Therefore, language is key.

While this analysis examines televised representations of Project Apollo from 1968 to 2022, it is first necessary to give a brief overview of NASA and its media policies. The extent to which NASA's Public Affairs Office influenced the networks' content will be evaluated. Knowing how NASA conducted its public relations will provide a broader picture of how the media reported and continues to report stories of space exploration. Although the story of NASA's public relations is provided in chapter 3, it should be noted here that information was gathered using an oral history approach. While rhetorical analyses enabled the author to interpret patterns and meanings of metaphors, conducting oral histories of NASA's original Public Affairs Officers offers clues as to where and how these metaphors originated when the Apollo program was conceived in 1961. Several of its original Public Affairs Officers, still living at the time that initial information was collected for this manuscript, shared their experiences, which have helped shed light on NASA's philosophy of space exploration and the ideology of the networks in their coverage of space exploration. This is important because NASA oftentimes supplied the networks with material to broadcast.

An oral history approach works nicely for examining this kind of relationship using Elizabeth Tonkin's (1992) rationale:

> Individuals are also social beings, formed in interaction, reproducing, and also altering the societies of which they are members. "The past" is not only a resource to deploy, to support a case or assert a social claim, it also enters memory in different ways and helps to structure it. We are our memories . . . we draw a social portrait, a model which is a reference list of what to follow and what to avoid.[13]

Following in the collective memory tradition, then, oral history helps to shape our future based on lived experiences. Along these lines, others such as James Hoopes have defined oral history as "firsthand collection and verification of any individual's spoken memory"[14] and historical data of individuals "who have first-hand knowledge about their life experiences, places that they have seen, people that they have met, and events that they have witnessed," according to Warren Nishimoto.[15]

Although there are numerous historical accounts provided by NASA Public Affairs Officers during the Apollo years, they tend to focus on interesting scientific or personal facts and have little to do with NASA's relationship with television networks. Therefore, interviewing retired Public Affairs Officers will help to explain this side of the "PR" dimension. These individuals were also selected because of their high rank in the Public Affairs Office and, thus, their authority to make decisions regarding what the networks were "given" and what might have been kept internally within NASA.

However, it should be noted that, while oral histories can provide excellent information, there are drawbacks. For example, David Henige warns that older participants may remember specific times of change as more "harmful than beneficial" because they focus on the time the change occurred rather than the change itself.[16] Thus some participants may distort experiences, remembering the past as a "metaphor" for their own lives. In addition, Henige warns that particular facts and events may be recalled, but the attitudes toward the events may be forgotten and replaced by new viewpoints, as each time something is recalled, modifications are more likely to occur.[17] To offset such possible distortions, multiple angles of the story are included despite recounting seemingly faded or contradictory memories. For example, if two retired Public Affairs Officers recalled an event differently, both sets of memories are included to offer a more complete story. But again, the purpose here is not to pinpoint "facts," but rather to gauge how narratives and metaphors have changed. Thus, knowing how the networks were given information from NASA in the first place might help to explain such changes.

Chapters 5, 6, 7, and 8 examine how Apollo was presented from 1968 to 2022. First is an analysis of Apollo news when it occurred, 1968–1972, on the *CBS Evening News*, focusing on what narratives—including storylines and metaphors—were used to communicate the complexity of the Apollo program. Again, CBS was the chosen network because it was number one in ratings in the late 1960s and early 1970s, surpassing NBC in 1966.[18] The first successful Apollo mission, Apollo 7, began in 1968 and the last mission, Apollo 17, ended in 1972. To view the news reports, a search using the word "Apollo" was utilized at the Vanderbilt Television News Archive (VTNA), using the VTNA online search engine. The search yielded a total of 340 stories pertaining to Apollo between September 1968 and December 1972, ranging from 10 seconds to approximately 20 minutes in duration.

In viewing each of the 340 stories, Mark Byrnes's framework was borrowed, in which he analyzed NASA's rhetorical strategies from 1961 to

1994 based on NASA publications and public speeches presented by NASA officials to shape its image. Byrnes defined each strategy as the following: nationalism deals with issues of national pride and security; romanticism addresses issues of exploration, heroism, and emotional rewards, such as challenge, inspiration, adventure, imagination, spirituality, and satisfying curiosity; and pragmatism emphasizes the practical use of space exploration while stressing the importance of the scientific and technological benefits it produces.[19] In viewing the CBS narratives, however, an additional narrative was identified: a technology theme that embodied military metaphors while also recognizing the importance of human involvement in technology production.

While at first it seemed logical to explain the segments chronologically, 1968–1972, after viewing several segments, it was evident that the reporting of the missions overlapped in narrative structure. For example, Apollo 11 was explained in terms of all four themes: nationalism, romanticism, pragmatism, and technology, while Apollo 8 focused mainly on romanticism. Therefore, the results are presented in terms of the four themes rather than the time frame in which they appeared. A second rationale for presenting the findings by narrative theme is that identifying the rhetorical patterns used, what the metaphors represent or mean, and how such narratives have changed over time offer richer narratives and historical accounts than merely reporting the number of narratives or metaphors. Because the arguments in this book are based on the assumption that television followed a ritual model of communication to present the Apollo story (e.g., reality is produced and maintained using commonly held beliefs such as patriotism and symbols such as the American flag), only rhetoric pertaining to these four rhetorical strategies was analyzed. All other forms of communication contained in the Apollo reports fell into the category that Carey has termed the transmission view of communication. These would include stories that began with introductions such as, "At Houston, the Apollo 8 astronauts and 1,200 workers at Johnson Space Center and Kennedy Space Center will get flu shots" (CBS, 12/3/68), or "The Apollo astronauts practiced today in a space flight simulator" (CBS, 12/16/68). These narratives are not included here because they tell us little about the social context in which the Apollo mission occurred, nor do they create and maintain the cluster of images and ideologies necessary to convince viewers to support Apollo missions.

It should also be noted that special attention is devoted to the themes in which metaphors are presented because, as many theorists have noted, metaphors enable the abstract to be given concrete form,[20] and "provide frameworks for making sense of the unintelligible."[21] And because space exploration was still in its infancy when the Apollo program began, administrators needed to provide a framework for the abstract concept of humans in space and man on the Moon. Therefore, David Eason's definition of metaphor was the operative one: "a comparison based on the principle of similarity" that "makes that which is unknown identifiable by locating it within cultural categories."[22] Examples include Walter Cronkite's describing the astronauts as "tourists visiting a foreign land," or the Saturn V rocket that took men to the Moon as "more powerful than a nuclear warhead," or the Moonwalk as "our last chance for world peace." Metaphors can be visual as well as spoken, and each broadcast was also analyzed for visual metaphors—such as the image of the American flag being planted on the Moon, or Apollo 8's "Earthrise" image—all of which connect the Space Race with our national interest.

Of equal importance is the degree to which the storylines used to describe the Apollo program have persisted, in some cases for more than half a century. Chapters 5, 6, 7, and 8 also address the memory of Apollo: how have the narratives used to describe the Apollo project changed, or have they remained the same? Which of the Apollo stories are remembered, and which have been forgotten? In keeping with Condit and others who have insisted that a broad range of discourse be analyzed to obtain a full measure of meaning,[23] four varied sites are considered: network news—Columbia Broadcasting System (CBS), cable news—Cable News Network (CNN), a cable entertainment network—The History Channel (THC), and a public television network—Public Broadcasting System (PBS). Examining the commemorations offered through network and cable news, documentaries offered on an entertainment channel, and educational programs offered through a public broadcasting network should satisfy the need for a broad range of discourse.

VTNA's online database was searched, and it yielded 124 Apollo stories that were broadcast on the *CBS Evening News* in the post-Apollo years, January 1973 through December 2022, which marked the fiftieth anniversary of Apollo's end in 1972. Of these 124 newscasts, 14 were excluded from analysis because they did not pertain to the Apollo space missions.[24] Apollo broadcasts ranged from 20 seconds to just over nine minutes in duration. For CNN, a search using the VTNA yielded 38 matches for Apollo. Because six unrelated newscasts were omitted,[25] there remained a total of 32 CNN news broadcasts pertaining to the Apollo space missions from 1994 to 2022.[26] These broadcasts ranged from 20 seconds in duration to 31 minutes.

However, because VTNA does not archive material for THC, a search was conducted on THC's official website, http://www.historychannel.com, now http://www.history.com,[27] to determine how many broadcasts pertained to Apollo. The word "Apollo" in THC's online search engine yielded 17 results. This number was misleading as the search engine provided every broadcast in which mentions of the Apollo program occurred. For example, one of the results was a show called *Superstitions* that dwelled on the origins of the superstitious number 13, and the extent to which the Apollo 13 mission was ill-fated merely because of its "unlucky" mission number. Because these types of programs tell us little about the Apollo program itself, only shows that were specifically related and limited to the Apollo program were included. Thus, once trivial or irrelevant shows were omitted, a total of seven shows would be analyzed, each of them 50 minutes in duration (commercials were omitted).

PBS documentaries are not archived by VTNA either. Therefore, a similar search was conducted on PBS's website, https://www.pbs.org/, using "Apollo NOVA" as the keyword, as the NOVA series focused on stories of science and technology. As with the THC broadcasts, shows that did not focus on the Apollo program were eliminated. Examples include shows that mentioned Apollo in terms of "post-Apollo" or those that briefly compared the Apollo project with other NASA endeavors such as Skylab. Not counting advertisements, trailers, or shorts that ranged from one to seven minutes, the search yielded four documentaries on Apollo, which ranged from 53 to 120 minutes in duration. A listing of all the broadcasts analyzed is included in the appendix.

Each newscast devoted to Apollo that appeared on CBS and CNN and each Apollo documentary or show that appeared on THC and PBS was treated as one broadcast, regardless of duration. Within each of these broadcasts, Eason's definition of metaphor supplied the criterion for identifying metaphors. Where applicable, a brief description of the narrative is provided so that the reader can follow the story lines. Narratives, then, were analyzed in terms of the rhetorical strategies—nationalism, romanticism, pragmatism, and technology in chapters 5, 6, 7, and 8.

Again, it is through language that meaning about a particular event is produced, maintained, and transformed. Project Apollo began as a goal: U.S. supremacy through space exploration. After the successful Moon landing, the Space Race ended. Thus, in the late 1960s and early 1970s, metaphors of scientific discoveries sustained the Apollo program and enabled NASA to branch out into the shuttle program to keep a human presence in space. In determining how Apollo is remembered and discussed today, language is the key unit of analysis: its codes of metaphors, values, and storylines enable culture to share symbols. Language does not give commands, but rather "seduces, offers, and persuades."[28] Knowing how the narratives in media representations have changed over time indicates not only what has transpired in science and technology, but also offers clues to the culture in which science and technology operate.

NOTES

1. CBS, *10:56:20 PM EST, 7/20/69: The Historic Conquest of the Moon as Reported to the American People by CBS News Over the CBS Television Network* (New York: Columbia Broadcasting System, 1970), 14.
2. David L. Eason, "Telling Stories and Making Sense," *Journal of Popular Culture* 15 (1981): 125.
3. See, for example, Michael Schudson, *Watergate in American Memory: What we Remember, Forget, and Reconstruct about the Past* (New York: Basic Books, 1992), and Barbie Zelizer, *Covering the Body: The Kennedy Assassination, the Media, and the Shaping of Collective Memory* (Chicago, IL: University of Chicago Press, 1992).
4. Gary R. Edgerton, "Television as Historian: A Different Kind of History Altogether," in Gary R. Edgerton and Peter C. Rollins, eds., *Television Histories: Shaping Collective Memory in the Media Age* (Lexington, KY: University of Kentucky Press, 2001), 10.
5. Alan J. Bush and Gregory W. Boller, "Rethinking the Role of Television Advertising During Health Crises: A Rhetorical Analysis of the Federal AIDS Campaigns," *Journal of Advertising* 20 (1991): 31.
6. Ibid.
7. Marouf Hasian and Emily Plec, "The Cultural, Legal, and Scientific Arguments in the Human Genome Diversity Debate," *The Howard Journal of Communications* 13 (2002): 304.
8. Horace M. Newcomb and Paul M. Hirsch, "Television as a Cultural Forum: Implications for Research," in Willard D. Rowland, Jr. and Bruce Watkins, eds., *Interpreting Television: Current Research Perspectives* (Beverly Hills, CA: Sage Publications, 1984), 58.
9. Steve Anderson, "History TV and Popular Memory," in Gary R. Edgerton and Peter C. Rollins, eds., *Television Histories: Shaping Collective Memory in the Media Age* (Lexington, KY: University of Kentucky Press, 2001), 21.
10. Gertrude Joch Robinson, "Television News and the Claim to Facticity," in Willard D. Rowland, Jr. and Bruce Watkins, eds., *Interpreting Television: Current Research Perspectives* (Beverly Hills, CA: Sage Publications, 1984), 202.
11. Herbert Gans, *Deciding What's News: A Study of CBS Evening News, NBC Nightly News, Newsweek and Time* (New York: Vintage Books, 1979), 147.
12. Raymond Williams, *Television and Technological Form* (London: Collins and Sons, 1974).
13. Elizabeth Tonkin, *Narrating Our Pasts: The Social Construction of Oral History* (Cambridge: Cambridge University Press, 1992), 4.
14. James Hoopes, *Oral History: An Introduction for Students* (Chapel Hill, NC: University of North Carolina Press, 1979), 7.
15. Warren Nishimoto, "Interviewing in Oral History and Journalism," *Oral History Recorder,* Summer, 1992, 2.
16. David P. Henige, *Oral Historiography* (New York: Longman, Inc., 1982), 46.
17. Ibid., 111.

18. Steven D. Stark, *Glued to the Set: The 60 Television Shows and Events that Made Us Who We Are Today* (New York: The Free Press, 1997), 126.

19. Mark E. Byrnes, *Politics and Space: Image Making by NASA* (Westport, CT: Praeger Publishers, 1994).

20. Claude Levi-Strauss, *The Raw and the Cooked: Introduction to a Science of Mythology*, Vol. I, trans. John and Doreen Weightman (Chicago, IL: University of Chicago Press, 1969, 1983), 341.

21. Roger Silverstone, *The Message of Television: Myth and Narrative in Contemporary Culture* (London: Heinemann, 1981), 10.

22. Eason, "Telling Stories," 126.

23. Celeste Condit, *The Meaning of the Gene: Public Debates about Human Heredity* (Madison, WI: The University of Wisconsin Press, 1999).

24. The excluded newscasts included Apollo Financial Services and stories pertaining to the Apollo Theater in New York City.

25. The omitted stories pertained to the Apollo Theater in New York City.

26. Although CNN began broadcasting in 1980, VTNA did not begin archiving its newscasts until 1994.

27. Due to a rebranding effort, The History Channel (THC) became History.com in 2008. Given the digital format offered on History.com's website and therefore hundreds of mentions of Apollo, many of which are articles or short one-minute videos, only THC full-length televised broadcasts were analyzed for this book.

28. Condit, *The Meaning of the Gene*, 253.

Chapter 3

NASA's Public Affairs Office

The Apollo Years

Although "the government has no little red book that spells out dos and don'ts for the public relations profession,"[1] Apollo 11 has been called a "public relations dream" and a "reporter's dream."[2] Indeed, there have been few events in history—especially positive ones—that have drawn the national and worldwide audience that the Apollo missions have. With the exception of the Manhattan Project, big science and public relations have traditionally gone hand-in-hand. As the adage goes, "big science costs big dollars," and scientific organizations need public relations to sell the importance of their projects, especially when the public is funding the endeavors. This was especially true for NASA during the Apollo years because it was the largest peacetime project ever undertaken by the United States government, and "probably rivals any technological undertaking at any point in time."[3]

While the Mercury and Gemini projects were justified to the public as a step toward beating the Soviets in the Space Race during the height of the Cold War, Project Apollo continued to emerge as the United States was experiencing a host of social problems: the Vietnam War, inflation, riots, campus unrest, civil rights, and women's rights issues. For many, tax dollars would have been better appropriated to fight injustices on Earth rather than being spent on an ideological quest in space. This was especially the case with sectors of opinion after the climax of Apollo's original goal: Armstrong's first lunar walk on July 20, 1969. Public support for space exploration began to take a steady and sharp decline. It was revived a bit during the Apollo 13 crisis because of public concern for the safety of the three Apollo 13 astronauts after the explosion in space aboard the spacecraft Intrepid. However, once the astronauts returned to Earth safely, public support again waned until Apollo 17, and the last three missions, Apollo 18, Apollo 19, and Apollo 20, were canceled due to a lack of Congressional support.

Although many observers have speculated that NASA made a conscious decision to shift Apollo mission rhetoric from "beating the Russians" to "scientific discovery" after the Apollo 11 lunar landing to keep the public captivated, NASA Public Affairs Officers, even 50 years later, affirmed that their role was to inform the public, not to persuade it. Conflicting reports raise questions about the openness of NASA, as will be evident later in this chapter. Such issues are important because public relations involve not just persuasive language but also accuracy. While the bulk of this research examines television narratives, it is imperative to examine the public relations mission of NASA, for it is through NASA that television news, especially in its early years, obtained information regarding Apollo. Thus, this chapter addresses how NASA's Public Affairs Office framed Project Apollo from the beginning to create and maintain public support.

NASA AND ITS PUBLIC AFFAIRS OFFICE (PAO)

The Origins of NASA's PAO: Pre-Apollo

After World War II, many federal agencies encouraged massive growth in public relations.[4] Just before the National Advisory Committee for Aeronautics (NACA) became NASA in 1958, the field of public relations underwent rapid change. In the mid-1950s, theorists argued that public relations practitioners should protect and promote their clients through a cautious dispersal of information. Simultaneously, however, journalists argued that PR practitioners, like journalists themselves, had a social responsibility to ensure the public's "right to know."[5] Thus, the tensions between journalists and PR practitioners, especially government PR practitioners, preceded the creation of NASA. NACA, created in 1920, understood that its mission was to organize and provide aeronautical research that would bring the United States to the same level of European aviation capabilities. Not until 1949 did NACA recognize a need to become more widely known to the public. Until then, "only the [aircraft and military] designers who depended so heavily upon NACA data seemed to know much about the agency."[6] Walter Bonney, an Illinois newspaper reporter and editor, became the head of NACA's Public Affairs Office in 1951 but often encountered reticence on the part of the military, which oversaw space operations. Before NACA became NASA, the U.S. Navy carried out the scientific research for the Vanguard rockets, and the Air Force developed ICBMs (intercontinental ballistic missiles). Their negative attitudes toward the press were similar to those of German rocket scientist Wernher von Braun, who was working on the Jupiter and Redstone rockets at the Ballistic Missile Agency in Huntsville, Alabama. Von Braun distrusted the press because he believed many journalists sensationalized and criticized his ideas.[7]

Like many organizations, the Army, Navy, and Air Force viewed press members as intruders. Not only were they concerned about top secret space-related information being leaked to "the competition," but they were also worried that news of failures would push the United States even further behind the Soviets in the public eye. Every botched rocket launch reinforced the military's belief that reporters should be banned from the Cape. The paranoia was infectious. When some reporters rented beach houses overlooking launch pads, neighbors, believing journalists were members of a Soviet espionage ring, tried to stop filming.[8] U.S. military helicopters would fly over camera crews to stir up sand to prevent crews from filming a launch.[9]

Walter Cronkite remembered using inference to obtain information that was not otherwise forthcoming: "We hung around [Cocoa Beach, Florida] bars, and when the engineers all disappeared from the bars one evening, we decided that there must be a shot about to take place. This was our source of our information."[10] The reporters and cameramen then adjourned to the Port Canaveral canal and waited on its rocks for hours for a rocket launch.[11] Before long the military realized there was no way to suppress launch times. Therefore, the Air Force thought it might be better to let journalists in and monitor what they did rather than have no control at all. As a result, camera crews were transported by the Air Force to observation areas on base to film a launch. The Air Force publicists then wrote a press release, and with Washington's approval, released it and the film one minute after launch.[12]

On October 1, 1958, NACA formally became NASA, and Bonney was appointed the first Public Affairs Director by NASA Administrator T. Keith Glennan. At this time, the Army's Jet Propulsion Laboratory and the Army's Redstone Arsenal, which housed Wernher von Braun's team of scientists who would build the rockets to send Americans into space, also became a part of NASA. In addition to creating an answer to Russia's Sputnik, NASA would oversee and coordinate a U.S. space effort previously distributed among the Army, Navy, and Air Force.

As a civilian agency no longer under the auspices of the U.S. military, NASA henceforth would make its goals, missions, and the research derived from those missions available to the public, as set forth in the Space Act of 1958. "Functions of the Administration" contained in Section 203 (a) (3) of the Space Act of 1958 established that NASA "shall provide for the widest practicable and appropriate dissemination of information concerning its activities and the results thereof," while Section 303 stated that information "shall be made available for public inspection, except (a) information authorized or required by Federal statute to be withheld, and (b) information classified to protect the national security."[13] In a memo that reiterated key sections of the Space Act, Bonney said bluntly, "Once NASA has done something, whether success or failure, it becomes something to be reported accurately and promptly."[14] By contrast, the same memo pointed out, "The Russians are carrying on their experiments in privacy, saying nothing about their failures, until they are successful."[15] Thus, not only was NASA to disclose its information by law, but also to set itself apart from the Russian space agency by being transparent.

Furthermore, Bonney outlined two goals for the new agency: one was for NASA to obtain a broader acceptance within the non-governmental or non-military service industry. The second was more long term: he envisioned a greater public recognition and understanding of NASA and its missions.[16] After the Space Act was signed, Bonney immediately began to devise a plan by which NASA's PAO, officially formed in December 1958, would operate. He believed that NASA "must maintain a positive information program designed to provide the people of the United States with maximum information about the agency's accomplishment."[17] Not only would the PAO provide information to Americans, but also to an international audience in an effort to dispel "the Communist lie."[18]

DEALING WITH THE PRESS

The memo from Bonney that outlined NASA's goals also included a section specifically devoted to the press:

> In servicing the press, the PIO [Public Information Office][19] seeks to function as a precision-ground mirror, faithfully reflecting the activities of NASA. In practice, PIO staff works as reporters within the agency, seeking out newsworthy information from NASA technical personnel and processing it into form useful to the press. The press uses this product of the PIO—the releases, the pictures—much as it uses the product of the wire services, with this important difference. It rewrites the product of the PIO and in the doing, makes the product its own.
>
> A large part of the PIO effort to serve the press is on a "response to query" basis. By providing prompt and accurate answers to questions, the PIO not only serves the press, but it relieves the NASA technical staff of much of the time-consuming chore of dealing directly with reporters.[20]

And journalists agreed that NASA did indeed serve the press. When NASA became a civilian agency in 1958, the press was welcomed, according to Walter Cronkite.

> When those fellows at NASA took over, they quickly realized that if the space program—the civilian space program—was going to [get] the billions of dollars through Congress, it had to have the public's approval, and therefore, public relations became part of the game.[21]

Many attributed this change to television, for it was this medium that brought stories of space exploration to the public. Even so, resistance in some quarters of NASA was high. Recalling visions of the U.S. Vanguard rocket exploding after the Soviets successfully launched Sputnik, NASA engineers were adamantly opposed to televising Alan Shepard's launch live.[22]

Despite concern for the public display of failure, President John F. Kennedy believed launches should be televised: the greater the risk, the greater the gain when a launch was successful.[23] In 1961, the Director of the U.S. Information Agency, Edward R. Murrow, asked NASA Administrator James E. Webb if manned spacecraft could be equipped with television cameras. Webb's Director of the Mercury Program, Abraham Silverstein, responded in a memo that the use of television cameras would have to wait for design and testing as the engineers did not believe that the Mercury booster capability would allow the extra weight of cameras, nor did they see any scientific value in having the cameras aboard. Astronauts also opposed having cameras aboard at first because they thought that operating the cameras would interfere with guiding the spacecraft.[24] As Walter Cronkite indicated during the Apollo years,

> There are those cynical enough—and I am one of them—to suggest that increased television from outer space is a kind of NASA ploy to see that the flight program continues beyond its scheduled flights. But there is more to it than that. I think that the astronauts understand now, as they did not always understand, an obligation to share their view of the universe with those of us who are earthbound. I think they want to show us tomorrow.[25]

Following the Mercury missions that flew between 1961 and 1963, the Gemini program (1965–1966) enabled space exploration to become part of the public's agenda.[26] By then, NASA had established its PAO, and television was given the green light. On June 3, 1965, it captured Edward White's first U.S. spacewalk on Gemini IV, which lasted twice as long as Soviet cosmonaut Alexei Leonov's had three months earlier. Ninety-two percent of U.S. homes equipped with a television set were tuned in to White's spacewalk, only two percentage points less than those for the Apollo 11 Moonwalk four years later.[27] Networks worked with NASA to provide live coverage of the Gemini X splashdown in July 1966. Shooting live undoubtedly augmented appeal.

Said Julian Scheer, Director of NASA Public Affairs during the Apollo era, "Everything that happened in orbit from all the NASA missions—I went to work for NASA at the last Mercury flight, from Mercury all the way through Gemini and through all the Apollo missions—[was] released in real-time without editing."[28] Thus, the Apollo missions were shot live, including the "The Big Show" from the Moon. Some at NASA, especially engineers concerned for logistical reasons, continued to oppose cameras onboard the spacecraft even for the Moon landing of Apollo 11. Scheer, the man credited with adding muscle to the PAO,[29] thought it would be "absolutely inconceivable, to have the capability of live pictures from the Moon and not do it."[30] Scheer prevailed. NASA asked RCA to design a camera to take to the Moon. The contract called for redesigning a 400-pound camera into one weighing a mere seven pounds. That camera brought NASA and the networks an audience of nearly 600 million people worldwide. Ironically, while we remember the images today as blurry, media reports during the time called the picture and sound quality "truly remarkable, considering the distance."[31]

Don Hewitt, Executive Producer of the *CBS Evening News* during the Apollo years and the creator of *60 Minutes* in 1968, described why he thought televising space missions was important:

> Nobody ever said it because nobody had to say it, but I always figured that there was an understanding between television and NASA—never spelled out, never even whispered, never even hinted at, but they knew and we knew. If we continued to help the space agency get its appropriations from Congress, they would in turn give us, free of charge, the most spectacular television shows anyone had ever seen.[32]

Similarly, Associated Press reporter Howard Benedict stated, "This new civilian agency brought a refreshing sense of openness to press coverage, as dictated by its charter. Until then, we'd had some difficulty explaining to the public the meaning of each failure or success."[33] Cynics, however, thought that NASA sought only benefits for itself.[34] Christopher Glenn of *CBS Radio*, for instance, believed NASA was extremely cooperative, but only to a point.

> There's something about NASA folks: they'll tell you anything you want to know but they won't tell you anything. They're so nice and so cooperative, but if you throw a curveball question, they'll get around it—very pleasantly, very nicely, but it's difficult to get hard information from them.[35]

Famed *Life* and *Time* photographer Ralph Morse was also ambivalent: "NASA wasn't always a help. There were some people in NASA who were excellent. But if you take NASA as a whole, I think they would have preferred it if we'd all stayed away."[36] Morse's response may have been colored by NASA's refusal to allow journalists to accompany astronauts on a survival training exercise in a Nevada desert.

Conversely, for the Apollo missions, NASA was open to the media. As Cronkite has noted,

> We were given absolute freedom to report the story, and a great flow of information from NASA. I would suggest that this freedom and the comparative reluctance of the Soviet Union to tell the facts until after the fact is an indication of the difference between an open and a closed society.[37]

When asked if NASA dictated to the networks what to report in Apollo news, one retired NASA public relations official (who asked to remain anonymous) replied,

NASA never could tell the networks or anybody else what to say or what to do. There were occasions, for example, during Apollo, when Julian [Scheer] wanted to make live communication between Apollo and Mission Control available to the press. As you can imagine, [he] was met with a great deal of resistance on the project management side of the house. But he won. And that stuff, as you know, was unfiltered. Now there was always private communications available to the medical people, between astronauts and medical people. Other than that, to the best of my knowledge, everything was open; it was live and in real time. We always did brag about the openness of the NASA programs.[38]

On very few rare occasions, some reporters threatened to sue NASA under the Freedom of Information Act (effective July 4, 1967), which forbids federal agencies to withhold disclosure of activities. This did not seem to bother NASA, which considered some of the requests from the press to be downright "ridiculous."[39] It should be noted that in 1971, NASA's acting administrator George Low tried to fire Scheer because Scheer "alienated some powerful members of Congress by refusing to cater to their every wish, such as providing special facilities for their friends during various lunar expeditions."[40] However, overall, members of the press held Julian Scheer in very high regard.

NASA'S "PR" FUNCTIONS: THE APOLLO YEARS

In a way, the Apollo program sold itself: NASA could not keep up with public inquiries from individuals and institutions (e.g., schools) that hungered for more information about Apollo and the space agency.[41] And NASA gladly released information knowing that public support and interest were a surefire way to keep funding abundant. But NASA had a plan for releasing information, too. In early 1968, the year the first successful Apollo mission would fly, Robert F. Allnutt, Assistant Administrator for NASA's Legislative Affairs office, wrote to the Honorable Charles E. Goodell in the U.S. House of Representatives that, while "NASA does not have what is commonly designated as a 'Public Relations' program," its objective of diffusing information about its activities "is met through a broad range of information and education functions." The letter specified several functions:

(1) Public information, which included handling daily requests, providing press site support, disseminating press kits and news releases in response to some 200,000 press and public inquiries per year.
(2) Special events, which included fulfilling requests for speakers, exhibits, overseas travel by astronauts, and awards ceremonies.
(3) Educational programs, which was part of the PAO in the Apollo days; it was responsible for relaying new space program information to elementary and secondary schools via teacher workshops, Spacemobiles (described below), publications, and films.
(4) Media development, which produced materials to distribute for public information, special events, and education functions of the PAO.

The letter claimed that NASA distributed over three million publications, mostly to school children who requested information, and that it loaned over 70,000 motion picture prints to the media, scheduled 3,800 speakers, created 1,000 exhibits, and conducted 11,400 Spacemobile lectures. Costs of these activities totaled $10.3 million in 1967.[42] By 1971, close to Apollo's end, NASA, one of the smallest federal agencies at the time, ranked third in public affairs efforts with 358 employees (1.3 percent of its total manpower) and annual activities costing $12 million (less than one percent of the total agency budget).[43] Clearly, NASA acted upon its mandate to inform the public and the press.

Like any organization, NASA issued press releases to keep the public informed, which were an especially important tools for NASA because of the highly technical nature of its missions. The agency used arcane terms, especially acronyms, as well as scientific language that had to be translated. While hundreds of press releases were issued during the Apollo years (ranging from changes in a mission's astronaut crew to announcements of new hires at NASA field centers), of more concern for the discussion at hand are the press kits specific to each Apollo mission that were distributed to the press corps, to other government agencies, and to the public. Because television networks planned their daily schedules around launch, landing, and splashdown times, the press kits became absolute necessities for production managers.[44]

In addition to NASA film footage, still photos, and rocket or spacecraft models, production managers used information from press kits to fill airtime during countdown delays or delays in splashdown. Press kits included material such as astronaut backgrounds, contractor data, and specifications for the mighty Saturn V rocket, often called the workhorse of the program. Most of the early press releases for the Apollo program (Apollos 7–10) were mainly technical and short. After Apollo 10, the press releases became longer and much more detailed in terms of purposes and procedures. Apollo press kits contained what NASA called a "General Release." It specified the purpose of the mission, briefly described mission objectives, named the astronauts and backup crew, provided a timeline of events, and was designed to give a brief overview of the mission. Subsequent sections offered heavier background or technical information, explaining "translunar injection," "transearth injection," fuel weight, biomedical monitoring, and solar wind composition, for examples.

The general releases did indeed push the scientific and technical aspects of the missions, more so with each release. A few, however, provided more than technical data. The Apollo 8 release justified the timing of the Apollo 8 launch date of December 21, 1968, on the grounds that the date fell within the window of opportunity for launching conditions because the Moon's position offered lunar surface lighting for when the spacecraft arrived at the Moon.[45] Those who were suspicious nonetheless concluded that NASA purposely scheduled that flight to arrive at the Moon on Christmas Day. The astronauts of this crew read the creation verses of Genesis from the Bible as they orbited the Moon on Christmas Day, 1968. Later, Madalyn Murray O'Hair sued NASA, claiming that the astronauts had violated the separation of church and state section of the First Amendment to the U.S. Constitution.[46]

Symbolism played an effective role in the press releases for the Apollo 11 mission, the mission that would see President Kennedy's goal obtained. The first page of the release reminds readers of man's "long-time dream of walking on another celestial body."[47] Next, it reproduced the plaque that was signed by the Apollo 11 astronauts and President Nixon and would be left on the Moon:

HERE MEN FROM THE PLANET EARTH
FIRST SET FOOT UPON THE MOON
JULY 1969 A.D.
WE CAME IN PEACE FOR ALL MANKIND

At this point, neither "race" nor the "Soviet Union" was mentioned, only "peace for all mankind."

More obviously symbolic was the foregrounding of flags that would be placed on the Moon: two large American flags, flags of the 50 states, the District of Columbia, and U.S. territories, flags of other nations, and one for the United Nations (UN). In addition, in every picture that featured an astronaut, a patch of the United States flag was clearly visible on the astronaut's spacesuit sleeve.

A third symbol was also evident at the beginning of the general release: President John F. Kennedy's name and a reminder that, "Successful fulfillment of this objective [lunar landing and return] will meet a national goal of this decade, as set by President Kennedy May 25, 1961."[48] As will be evident later in the discussions of televised representations, the Apollo story could not be told without the American flag or an icon of Kennedy.

A final feature regarding the Apollo 11 general release highlighted the fact that during the Apollo years, Armstrong was a civilian while the other two crew members, Aldrin and Collins, were ranked officers of the United States Air Force. This is significant because, by the time the Americans were to land on the Moon, NASA wanted the mission to be a symbol of peace for all mankind instead of a feat against the rival Soviets. It was rumored for many years that Deke Slayton, in charge of the astronaut corps during the Apollo years, handpicked Neil Armstrong to be the first American on the Moon because he was a civilian and because NASA wanted to avoid any military connotations of the United States "conquering" or "claiming" the Moon as its territory. While NASA Public Affairs Office Director Julian Scheer openly and strongly suggested the first astronaut on the Moon be a civilian for this reason, it was Slayton who maintained a systematic crew rotation so as not to anoint a candidate prematurely, and thus to maintain morale among the cadre of astronauts.[49]

Toward the end of the program, the general releases thinned to specific information, perhaps because the latter missions were continuations of previous ones or, alternatively, because there was less of a need to sell the program.

Spacemobiles

The Aerospace Education Services Program, or "Spacemobile," began in Washington, D.C., in 1961 as a way to educate students and teachers and pique their interest in space science. Spacemobiles were large vans containing 1,000 to 1,500 pounds of space-related equipment to demonstrate principles of flight such as gravity and orbital mechanics. Though it was scaled down and later changed its name, the Spacemobile educated students of all ages during its existence. The last mention found of the Spacemobile in the NASA History Office archive is in 1998. But in its heyday during the Apollo years, Spacemobile educators were required to have a bachelor's degree and be a teacher or an administrator. Most were between the ages of 25 and 35. On rare occasions, educators received visits from astronauts as part of their training for the job of "Spacemobiler." As with the astronaut program, NASA received more applications than it needed for Spacemobile educator positions. Once a Spacemobiler was hired, NASA was especially concerned with his or her performance. Although these instructors were contract employees not directly employed by NASA, the final approval for employment rested with NASA, and it was NASA to whom they reported.

In addition to live "gee-whiz" presentations and student participation in aerospace activities, such as trying on spacesuits and participating in exercises to illustrate centrifugal force laws, Spacemobile activities for students ranged from watching photographic slides and videos to receiving literatures on the space program. Spacemobiles were so popular that, in some cases, schools waited three years to receive a visit. One student, Jerome Apt, was inspired by a Spacemobile program in the early 1960s to study math and science. After graduating from Harvard magna cum laude, Apt later received a PhD in physics from MIT, then became a Shuttle astronaut and physicist. Naturally, the Spacemobile program held Apt up as an example. A NASA press release describing the program made an additional appeal to students: "Today's students who enter the math, science, and technology pipeline will be the generation who transforms America's great new adventure in space exploration into a reality. They will be the first colonists of the Moon and the first Martians."[50] Using rhetoric such as "adventure" and "first colonies" surely conjured up images of Columbus's adventures to the New World and frontierism. It is no surprise, then, that the Spacemobile program officially began on May 5, 1961, the same day that Alan Shepard became the first American in space, and 20 days before President John F. Kennedy made his famous speech committing America to land a man on the Moon before 1970. Having made his promise based on NASA's 15 minutes total experience in space, Kennedy knew that the country needed the skills of mathematicians and scientists to help it achieve that goal—a goal that even NASA engineers doubted could be accomplished in nine years.[51]

The President also used the Spacemobile to persuade foreign governments of the merits of space exploration. For example, when the Venezuelan president visited the White House in the spring of 1963, Kennedy wanted to give him something tangible regarding space exploration to take back to the people of Venezuela. Kennedy's press secretary, Pierre Salinger, who happened to be good friends with Bill Lloyd, NASA's Public Affairs Director, suggested Kennedy send a Spacemobile to Venezuela. This sparked an international Spacemobile program that in the end provided space exploration education to over 50 countries around the world, including "one country that was behind the Iron Curtain, Yugoslavia."[52] Requests for the Spacemobiles were routed through the USIA and the State Department, and information about the program was sent to U.S. embassies in countries that requested it. Because Spacemobile educators were sent overseas, they were monitored by NASA and the State Department. Since members of the State Department are not typically scientifically trained, the space agency arranged to meet with embassy personnel during normal briefings with the State Department to reassure them that instructors were following NASA-provided scripts.

Because President Kennedy personally supported the program, requests for Spacemobile artifacts were rarely denied.[53] However, funding was drastically cut in the mid-1960s, and the *international* Spacemobile program was eliminated in 1965, a victim of the Vietnam War and Johnson's Great Society domestic programs. Since Nixon's interest in the space program was entirely political, he did not do much to step up funding for the *domestic* Spacemobile program during his term in office. As a result, the program diminished in size and activities and eventually the program was transferred from NASA's Public Affairs Office to NASA's Office of Education to be managed by contractors who specialized in educational programs. In 1991, 35 professional aerospace instructors spent about six months of the year on the road, each traveling some 40,000 miles annually to American classrooms. In the 30 years between 1961 and 1991, the Spacemobile program educated more than 57 million students and more than 400,000 teachers in lectures and workshops.[54]

Image-control 101: Symbolism, Rhetorical Strategies, and Crisis Management

Hugh Harris, who retired in 1998 as Director of NASA's Public Affairs, was a Chief Public Information Officer at NASA during the Apollo years. When asked about the purpose of NASA's Public Affairs Office, he replied that NASA really was not in the business of "PR."

> The philosophy in NASA at the time [of Apollo] was basically that we did not do PR. Our job was to supply the facts to the media and to the general public through a number of avenues. But we didn't try to promote or go out and change opinion or have strategies to do that. The philosophy was that if you're telling a very straight forward [story] and why it's important, then the public is smart enough and the legislators and all that sort of thing to recognize whether it's something the country should support or not.[55]

Regardless of NASA's philosophy, the fact remains that there was a public relations sector of the agency; the PAO's directive from Congress was to inform the people. All federal agencies dependent on funding must create a positive image to survive.[56] Three ways are standard. The first is to create an image through symbolism—a kind of branding that enables audiences to identify with the agency. Take, for example, Nike's "swoosh" logo, a symbolic representation of the wing of the Greek goddess Nike, who ran and flew at great speeds.[57] A second is to "define" the organization through rhetorical strategies or through slogans (e.g., Dunkin' Donuts' "America runs on Dunkin"). A third is crisis management, how an agency conducts itself in the face of catastrophe. NASA's PAO depended heavily on all three of these methods.

SYMBOLISM

Branding Space Symbolically: Using "Old Glory"

Flags, or "emblems of faith," have always inspired and "serve[d] to publicize and prove the great achievements made by a particular nation."[58] Flags have also enabled people to display faith and love of country. For example, after the deadliest terrorist attacks in the United States on September 11, 2001, store owners could not keep American flags stocked because flag manufacturers could not produce them fast enough. Nearly 40 years prior, strengthening the image of astronauts as all-American patriotic men, John Glenn affirmed to a joint session of Congress after his historic Earth orbit, "I still get a hard-to-define feeling down inside when the [American] flag goes by."[59] Using the flag as a symbol of success was especially prevalent during the Apollo years. NASA must have known that the sight of "Old Glory" implanted on the Moon would also pull at the heartstrings of the American public on the day it realized Kennedy's promise would ring true. Making that promise in 1961, Kennedy returned to it a year later: "For the eyes of the world now look into space, to the Moon and to the planets beyond, and we have vowed that we shall not see it governed by a hostile flag of conquest, but by a banner of freedom and peace."[60] Indeed, the American flag would become that banner.

From the beginning, NASA embroidered the Stars and Stripes as a patch on astronaut uniforms and stenciled it on launch vehicles and spacecraft in the Mercury and Gemini programs. But Apollo was different because it would culminate in a flag on a staff planted on the Moon. The decision to plant a flag on the lunar surface was controversial. The Outer Space Treaty signed by several member states of the United Nations, including the United States, declares, "Outer space, including the Moon and other celestial bodies, is not subject to national appropriation by claim of sovereignty, by means of use or occupation, or by any other means."[61] Legally, the United States could not claim the Moon, but symbolically it could, and it did. Placing the Star-Spangled Banner on the Moon would conjure up images of Christopher Columbus planting the flag of Spain in the New World. Any territorial claims would be offset by the plaque affixed to the Apollo 11 lunar module, to be left on the Moon, which read, "We came in peace for all mankind" and was signed by President Richard Nixon. Six months before the Apollo 11 flight, Nixon observed in his first inaugural address, "As we explore the reaches of space, let us go to the new worlds together—not as new worlds to be conquered, but as a new adventure to be shared."[62] Thus, while the President affirmed an international purpose, the flag would represent the United States politically.

NASA formed the "Committee on Symbolic Activities for the First Lunar Landing" in February 1969 to devise a way to "signalize the first lunar landing as an historic forward step of all mankind that has been accomplished by the United States" without implying that the United States was claiming the Moon as a territory and that would not jeopardize the astronauts' or the missions' safety.[63] As we know, the committee recommended that the United States flag and the plaque be left on the Moon: the former would represent the "first historic forward step" for the United States, while the latter would indicate "men from the planet Earth" to mean *all* men, not solely "men from the United States."

For Apollo, the flag symbolized America's pride: members of Congress agreed with NASA's Committee on Symbolic Activities that it should be displayed on the Moon. Senator Wallace F. Bennett (R-UT), for example, wrote to Administrator Thomas Paine, "Let's look out for ourselves for once. It is getting a bit tiresome to be kicked around for failure, so let's broadcast success."[64] Representative John R. Rarick (D-LA) concurred, "When our men land on the Moon, history and national pride demand that Old Glory be raised there."[65] NASA informed Congress on June 10, 1969, a little more than one month prior to the Apollo 11 launch date, that the astronauts would place the red, white, and blue on the Moon. An agreeable Congress later formally amended the Outer Space Treaty on November 4, 1969. The amendment is Section 9 of the NASA Authorization Act approved by the House of Representatives:

> The flag of the United States, and no other flag, shall be implanted or otherwise placed on the surface of the Moon, or on the surface of any planet, by the members of the rest of any spacecraft making a lunar or planetary landing as a part of a mission under the Apollo program or as a part of a mission under any subsequent program, the funds for which are provided entirely by the Government of the United States. This act is intended as a symbolic gesture of national pride in achievement and is not to be construed as a declaration of national appropriation by claims of sovereignty.[66]

Although members of Congress and the public argued against placing a United Nations flag on the Moon,[67] flags for member countries of the UN were flown aboard the Apollo mission and later presented to heads of state. Small flags for the District of Columbia, U.S. territories, and each of the 50 states were also flown to the Moon and later presented to each of the governors of those states.[68] While there were no formal public outcries, some were disappointed that the UN flag did not receive similar recognition, and still others believed the Christian flag should have accompanied the American flag.[69] Eventually, demand for Moon flags would soar: an American flag from the Apollo 12 crew was auctioned for $25,000, the winner outbidding John Wayne, a frontiersman of another sort, for the Stars and Stripes.[70]

The flag planted on the Moon by Armstrong and Aldrin was a 3′ by 5′ nylon banner attached to a gold-toned 8′ aluminum staff, supported with wire to look like it was waving in the absence of wind on the Moon. In a final symbolic gesture, President Nixon asked that the American flag be displayed on all public buildings on July 21, 1969, the "National Day of Participation," as set forth in Proclamation 3919:

> As the astronauts go where man has never gone, as they attempt what man has never tried, we on Earth will want, as one people, to be with them in spirit; to share the glory and the wonder, and to support them with prayers that all will go well.[71]

Upon returning to Earth, Apollo 11's Armstrong said, "To me it was the proudest moment of my life, to be able to stand there and quietly salute the flag."[72]. The flag, in short, came to mean Apollo. Even today, Americans associate the Moon with the American flag. In 1981, 12 years after the first Moon landing, Music Television (MTV) debuted, using as its logo the likeness of the famous shot of Aldrin in his spacesuit saluting the newly planted banner. Armstrong, however, refused to lend his voice intoning, "That's one small step for [a] man, one giant leap for mankind"[73] as the MTV flag was planted.[74] But even without the words, the image conveys MTV's message to cable viewers: "We've made it; we're here to stay."

BRANDING SPACE SYMBOLICALLY: MISSION PATCHES

To commemorate events, NASA designed and distributed emblems, or patches, to NASA employees and to the public for purchase in NASA Exchange stores and NASA visitor centers, as well as in science museum gift shops throughout the United States. The patches became a way for the public to identify each mission according to its specific goal. Thus, to know the history behind the patches would be helpful for the task at hand. Although the astronauts themselves designed patches for each of the specific Apollo missions, all designs required NASA approval before the "official" version was produced by the A.B. Emblem Company in Weaverville, North Carolina. To manage the patches and control how they were used on products, NASA prohibited the unauthorized manufacture and sale of official patches, a misdemeanor punishable by a fine of $250 and a prison term of six months. Some companies did legally produce and sell insignias with their own designs, but NASA's official patches contain the symbolism the agency wished to share with its mediated audiences. Patches were designed only for manned missions, beginning with Apollo 7. Previous Apollo missions were test missions and unmanned.

Patches reflected the military origins of the astronauts and the agency. Military patches primarily identify units. Durability, ease of use, and cost were factors. Most importantly, a "uniform" identifier encouraged "team thinking," a common strategy in the training that astronauts had received as test fighter pilots. While military patches ordinarily employ three to six colors for a design, NASA patches were more colorful, with 10–19 colors.[75] NASA wanted to identify itself as a civilian agency, albeit one with some military associations, so the patches satisfied employee and employer.[76] NASA discovered that the public liked them, too. "Patches" became popular in the 1960s when students and other groups applied emblems to jeans and jackets to signify their concerns and to promote causes.[77] The astronauts wore their mission patch, a U.S. flag patch, and the NASA patch on their spacesuits for each of the Apollo missions. Today, the mission patches are collectors' items.

Because patch designs symbolized specific elements, they helped the public keep in mind each mission's goals and ideals. With every mission, the patches steadily incorporated more meaningful symbols of American idealism, culminating in the complex design for the final Apollo 17 mission. Apollo mission patches 7 through 17 are shown in figure 3.1, to illustrate how the symbolic images mirrored the narratives that drove the mediation of Apollo.

Predictably, symbols of America were the most prevalent. For instance, Apollo 11, the most widely viewed and reported Apollo mission, and the one that fulfilled President Kennedy's goal, was commemorated by a design that featured an American Eagle in the center. The eagle, in shades of gold, stood out against the dark blue color of deep space. Although the eagle symbolized the American spirit, the astronauts wanted to emphasize that the mission was for all mankind, so they placed an olive branch, symbolic of peace, in the eagle's beak. The eagle appears to be flying in for a landing with its feet extended and its wings partially folded. NASA did not approve of the original design because they thought the talons looked too hostile and warlike. The olive branch was instead placed in the eagle's claw to soften the look. Rather than the astronauts' names, the original design featured "Apollo Eleven" at the top of the circular patch. Armstrong argued that those who did

Figure 3.1 Saturn Apollo Program Patches. ***Source*****: Courtesy of NASA. For information only; promotional usage is forbidden.**

not speak English might not understand "Eleven" so "11" was used instead. While not illustrated here, another example of Americanism is the ill-fated Apollo 1 patch, designed by the three Apollo 1 astronauts, Ed White, Gus Grissom, and Roger Chaffee, who died in the Apollo capsule fire on the launch pad in 1967. It featured each of their names around the module orbiting the Earth with the Moon visible in the background. The Moon is sewn in white with a silver ring trimming it, perhaps creating a halo effect.[78] In the background also waves an American flag, with the Stars and Stripes in red, white, and blue. Having acquired tragic resonance, this insignia was later flown to the Moon on the Apollo 11 mission in memory of White, Grissom, and Chaffee.

Patches also represented the astronauts' military affiliation. While NASA was careful not to incorporate militarism in its missions because it did not want to appear to claim the Moon as U.S. territory, the inference of the astronauts' military background reminded viewers of the astronauts' knowledge and training, and their service to the United States. The design of the Apollo

12 mission patch used the Navy's color scheme of blue and gold, symbolic of the three astronauts' Navy background. It also featured a ship, patterned after the U.S. Navy's "Yankee Clipper," landing on the Moon with the American flag flying at the top of the foremast. The astronauts believed that the Yankee Clipper "embodied the spirit of America, and the spirit of naval camaraderie," and it trails a rocket blast to symbolize the technological achievement of space exploration.[79]

Patriotic red, white, and blue emphasized the principle of teamwork in the Apollo 15 patch. It highlighted the collective efforts of NASA's employees and its contractors on behalf of America to anticipate the continuing missions. In the Apollo 16 patch, the Moon is the background for a bald eagle perched on a red, white, and blue shield. Through the shield runs a sort of wishbone, the same symbol used in the NASA logo to represent the agency and spaceflight.

The last of the Apollo missions, 17, mixes symbols of Greek mythology with ideological icons to "capture Man, Future, the Recollection of Apollo, the Flag and its contemporary heritage."[80] In the foreground and to the left of the patch is a depiction of the Greek god Apollo's head, symbolizing the Apollo program. Gold thread depicting Apollo in the foreground and the Moon in the upper right-hand corner evokes the golden age of space exploration. Apollo and the American eagle look forward toward depictions of Saturn and other planets and stars to the right, indicating that America has much to explore in the future. The eagle's American heritage is visible in the blue background and four red stripes and three white stars (representing the astronauts) that make up its body. Its wings span the lower part of the Moon to symbolically indicate that the United States has already been there.[81]

Also mythical was the design of the Apollo 13 patch. The astronauts sought the help of New York City artist Lumen Winter, who conceived the idea of having three horses pulling a chariot to symbolize the three astronauts and their spacecraft. The sunrays in the background symbolize Apollo pulling the sun across the sky every day. The Earth is sewn below the horses' hooves, implying that the Apollo chariot would once again reach heavenly orbit. Because public support for the program was waning, the patch also reminded the public of Apollo's scientific purpose. To the left of "Apollo XIII" in the uppermost part of the circular patch is "Ex luna, scientia," Latin for "From the Moon, knowledge," which might have been the astronauts' way of justifying yet another mission to the Moon.

Some of the patches also featured North and South America, the most prominent and recognizable land masses of their depiction of the Earth, as was evident in mission patches for Apollo 7, Apollo 8, and Apollo 10. With the letters "USA" vertically placed along the length of the rocket, the Apollo 9 patch reminded the public that the technology needed to get men to the Moon was made in the United States.

Telling stories of their own, then, the patches embodied the missions' goals and ideologies. As we can see, with every mission, the patches used more American symbolism to reinforce the necessity of space exploration even after it achieved a landing on the Moon. Thus, as will become even more evident later, the patches proved to be durable "brands." CBS constantly alternated between using the patches and pictures of the astronauts specific to each mission in its coverage, even in post-Apollo years when CBS commentators introduced the anniversary of a mission. Interestingly, the patches reappear in telecasts, such as on The History Channel, still branding missions decades later.

BRANDING SPACE SYMBOLICALLY: NAMING THE SPACECRAFT

Also symbolic were the names the astronauts chose for their spacecraft. Following U.S. military tradition, beginning with the Apollo 9 mission, the Apollo astronauts bestowed on the spacecraft they would fly names they thought either emphasized the purpose of the mission or, in some way, paid tribute to the country that sent them into space. Again, many of these names not only branded the mission but symbolically registered culturally identifiable objects to which the public, including children, could relate. For instance, astronauts picked names they thought the Apollo 9 spacecraft resembled, selecting "Gumdrop" for the command module and "Spider" for the lunar module. For Apollo 10, the lunar module was named "Snoopy" after Charles Schulz's cartoon character because the astronauts imagined it snooping around the lunar surface, and "Charlie Brown," Snoopy's companion and guardian, was given to the command module.[82]

There were also symbolic patriotic undertones. The Apollo 11 lunar module became "Eagle" in keeping with the regal bird in the patch. Julian Scheer in NASA Public Affairs suggested the astronauts name the command module "Columbia" for several reasons: it was similar to Jules Verne's "Columbiad," it recalled Columbus, and it was once suggested as the name for the United States.[83] Similarly, for the Apollo 12 mission, the astronauts named the command module "Yankee Clipper" because they believed it embodied "the spirit of America and the spirit of naval camaraderie."[84] The lunar module was the "Intrepid."

The names given to the craft of Apollo 13 proved appropriate after the explosion aboard the spacecraft. The command module was named "Odyssey," recalling Ulysses's "long voyage" of many perils before he managed to return to Ithaca alive. The lunar module was named "Aquarius" after the Egyptian god, not the popular late 1960s song by the Fifth Dimension in the musical "Hair" as many believed it had been. Aquarius symbolized a water carrier who brought fertility—life and knowledge—to her worshipers.[85]

To honor fellow Americans, the Apollo 14 astronauts named the command module "Kitty Hawk" in honor of the site in North Carolina where the Wright Brothers pioneered the first airplane. The lunar module was named "Antares" after the most visible star in the Scorpius constellation. Since its crewmembers were Air Force veterans, the lunar module for Apollo 15 was named "Falcon." NASA geologist Farouk El-Baz borrowed the name "Endeavor" for the command module from a children's story about a ship (named Endeavor) that took Captain Cook on scientific voyages to the Pacific. Patriotism aside, the astronauts wanted to emphasize Apollo's scientific rationale.

Similarly, consideration for younger Americans shaped choices for Apollo 16. Astronaut Ken Mattingly overheard a child saying that the astronauts on the Moon looked like Casper because their white and shapeless Teflon space suits resembled the "friendly ghost." Mattingly "liked the idea of something that was not so serious and which kids could identify with."[86] Thus, "Casper" became the command module, while "Orion" was given to the lunar module. Finally, as a way to recognize and thank the American public who funded the mission, the Apollo 17 crew named the command module "America" and the lunar module "Challenger."[87]

"ENGINEERING CONSENT": NASA'S RHETORICAL STRATEGIES

In *Politics and Space: Image Making by NASA*, Mark Byrnes argues that in constantly changing political environments, agencies purposefully recast their images to fit those environments. This becomes especially necessary to maintain funding for projects and to maintain the authority to run them. Again, in examining his study of the rhetoric used by NASA officials and discursive strategies used in NASA publications such as brochures, Byrnes identified three dominant themes used to build public and thus political support: nationalism, romanticism, and pragmatism. Each strategy accorded with the political necessity of the day. For instance, during Project Mercury, 1958–1963, NASA's first major program and the country's first manned program, the agency depended heavily on a rhetoric of nationalism emphasizing that the space program benefits America as a nation, especially in light of the Soviet threat. Thus, the emphasis on competition with the Soviets during the Cold War was dominant during the Mercury era. NASA gained mileage through constantly evoking the Soviet threat and the urgent need for American reaction and by appealing to national pride and national prestige. This strategy spoke directly to Americans' love of country.[88] An example appears in NASA Administrator T. Keith Glennan's 1959 statement to Congress:

> We are engaged in a *struggle* for the *minds and hearts of men everywhere*. The issue is simply whether *our* system of *free government* and *responsible civic freedom* is *superior* to the system of *totalitarian communism* and *forcible* direction of the lives of its *captive* peoples. I believe it is becoming increasingly obvious to the world that *Russia's* space activities are *devoted*, as are *most of their activities* as a nation, in large part to the *furthering of communism's unswerving designs upon mankind.*[89] (emphasis mine)

Even NASA's written objective incorporated the rhetoric of nationalism, "the establishment and maintenance of a *strong national capability* to operate in space and to use space fully in the *national interest*." This would give the nation "*freedom of choice* to carry out whatever missions the *national interest* may require—be they for *national prestige*, *military requirements*, scientific knowledge, or other purposes," and would also "*prevent any other power from denying us* the utilization of space in *our interests*"[90] (emphasis mine).

A second rhetorical strategy to maintain political support was the use of romanticism, which includes themes of exploration, inspiration, adventure, challenge, spirituality, and curiosity. NASA wanted the public to participate vicariously in astronauts' exciting and heroic activities, and it underlined the spectacular nature of its endeavors and thus offered the public a way to temporarily escape the humdrum routine of daily life. Events in the late 1960s made romanticism seem more appealing than nationalism. As the United States eroded the Soviet Union's lead in space, Americans felt less threatened by the Soviet space program and found nationalism less compelling. Apollo also offered an escape from a host of social problems in the United States, including the Vietnam War, social unrest, riots, race relations, poverty, women's rights, countercultural voices, and assassinations. NASA sold the Apollo program as the bright spot in a troubled society. Administrator James Fletcher appealed to Americans' aspirations when he spoke of a "*human hunger* for *positive elevating goals*."[91] The agency also reminded the public that it needed spirituality amid social ills, and it could provide that, too.

> The exploration of space can *give* you *new interests* and *new motivations* arising from an *expansion of your intellectual and spiritual horizons* as you take a longer view of *man's role in time and space* at this point in the history of the *human race*.[92]

In addition, the administrator used the frontier metaphor by stating that the United States "must be a country that continues to *look to frontiers* of science, *frontiers* of technology, *frontiers* of mankind."[93] Within the frontier narrative is adventure, which NASA veritably sought to persuade the public is innate in humans.

> We have just *opened the door* into the *limitless reaches* of the universe, and we can see just far enough ahead to know that man is at the *threshold of a momentous area.* Here is *opportunity, challenge, adventure so tremendous* to exceed anything which has gone before. Here is *the tomorrow* which *youth* wants to *embrace*, and which we must not deny because of a waning of the *frontier spirit* which *made America great.*[94]

NASA also cast its goals in the American imaginative tradition:

> *Imagination* has helped to *pave new paths* in technology and to *chart the course* of science. It has *nourished great art and literature* and *opened the way* for *civilizations to prosper.* And perhaps most important, it has intimately connected with the *notions of freedom* and *self-fulfillment* that we in the *democratic world hold so dear.*[95]

Later, traditional American pragmatism surfaced to insist that NASA was, above all, a practical scientific agency once the race to the Moon was won. That device continued to be useful during the Shuttle program. As NASA Administrator James Fletcher declared:

> We now *look ahead* to several decades of a *highly rational use* of space, and the turning rapidly *developing* space *capabilities* to *useful work.* We have made our new program *relevant* to the *needs of modern America.* We have turned from a period of space exploration to a period of *space exploitation for practical purposes.*[96]

Those purposes included remote sensing of the Earth, satellite communications, and manufacturing in space. Fletcher added, "By learning more about our planetary environment, we will be *better able to apply* the *knowledge we gain* in *practical ways* to the *enrichment of life on earth.*"[97]

Byrnes concluded his study by stating that, while NASA has largely been successful in attracting public support to secure political backing, the Challenger accident and constant trouble with the Hubble Space Telescope have made those tasks more difficult.[98] As will be evident below, the Apollo years also provided arduous PR tests for NASA's public affairs specialists, who learned difficult lessons in crisis management, but ultimately restored confidence in the Apollo Project and the space program in general.

PR'S TRUE TEST: CRISIS RESPONSE

Eleanor Roosevelt said that, like a tea bag, you can never tell how strong a woman is until she is in hot water. The same can be said for organizations' ability to handle adverse situations. NASA found itself in such a quandary when the Apollo program suffered two crises in its 12-year history: the Apollo 1 fire in 1967 that killed astronauts White, Grissom, and Chaffee on the launch pad, and the explosion in space aboard the Apollo 13 capsule in 1970 that almost claimed the lives of three additional astronauts. The first crisis NASA handled poorly, leaving Congress and the public wondering if the project was worth the three lives lost. The second mishap not only provided a model way of handling a crisis but also strengthened NASA's image and actually caused NASA to revive its public and Congressional support for future missions.[99] This section will examine the Apollo 1 and the Apollo 13 incidents to shed light on how NASA communicated with its public when things were not going so well and to illustrate that the way in which NASA "spun" the crises influenced the public and Congress—entities NASA is required to keep informed according the the 1958 Space Act.

When the fire sparked in the Apollo 1 capsule on January 27, 1967, the astronauts were unable to open the hatch, and their pleas for help were heard in the blockhouse by those monitoring the pretest flight, but could do nothing to save the astronauts' lives. NASA was shocked: how could this agency, the technological leader in innovation and engineering, have not foreseen and corrected faulty wiring in the capsule? And how could an organization that worked so hard to cultivate and manage an image of perfection with the Mercury and Gemini astronauts tarnish itself with such carelessness? The nation was also shocked that NASA could let this happen to three of its space heroes, three of its troops.

The fairly new NASA, only nine years old when the Apollo 1 fire occurred, scrambled unsuccessfully to minimize its loss. When a crisis occurs, public relations professionals are taught to "respond quickly; tell the truth; and provide a constant flow of information, especially to key publics," and NASA did none of these.[100] Even though its policy was to release an alert within 20 minutes of learning of the loss of life, NASA took two hours to report that White, Grissom, and Chaffee had been killed. Officials claimed that they had difficulty locating one of the astronauts' next-of-kin and wanted to keep the press from bombarding the families of the astronauts with questions and reactions. When NASA finally did respond, the reports were inaccurate and misleading: an initial press release indicated "a fatality" instead of "fatalities."[101] Agency officials also gave contradictory statements in reference to technical questions about the capsule. While no one thought the errors were malicious, skepticism grew among press members. NASA's imposing a news blackout made the media and government officials even more suspicious. Because they had no official information, the media reported half-true statements just to have something to say to the public.[102]

To worsen the situation, NASA appointed its own board to investigate the tragedy, a board comprised of individuals in the industry with close ties to agency officials. Administrator James Webb put together a panel he thought would best get NASA back on track. On the evening of the Apollo 1 fire, he met with President Lyndon B. Johnson, who agreed NASA should have its own board, which that same evening was created, and all information pertaining to the fire was impounded.[103] Only after public outcry and pressure from Congress and the media did NASA release information, and only after separate investigations by House and Senate boards did the public learn about the shoddy workmanship of the Apollo capsule. Its reputation tarnished and its Public Affairs Office distrusted, the space agency had to work toward a future mission that would restore its image.

That day came with Apollo 8, the mission that sent Lovell, Borman, and Anders around the Moon on Christmas Day. The reading of Genesis seemed to restore any doubt the public might have had; God seemed to be on NASA's side this time. While Apollo 6 and Apollo 7 were also successful, it was Apollo 8 that got closer to the Moon than any NASA mission ever before. With the successes of Apollos 9 and 10, and of course the lunar landing of Apollo 11 televised to the world, the public seemed to forgive and forget the Apollo 1 fire and basked in NASA's consecutive achievements. Success bred public complacency, however. Americans began to question why additional trips to the Moon were necessary. After all, the United States had won the space race, and now the country had other problems that demanded economic and social attention. Then, crisis ended complacency. On April 13, 1970, the world would hear the iconic phrase that has now become cliché, "Houston, we've had a problem."[104] An oxygen tank feeding the electrical system exploded when an Apollo 13 astronaut performed a routine stir of the liquid oxygen. The explosion triggered a short circuit that threatened life support systems aboard Apollo 13. With painful memories of the Apollo 1 fire resurfacing, the public once again became suddenly interested in the program, and stayed glued to the televised drama, hoping to hear that the astronauts would be returned safely to Earth.

At this juncture, NASA needed Congressional and public support, and could not afford another debacle in handling a crisis. Having learned a lesson from tragedy, the agency knew that suspicion of a cover-up would surely end the Apollo program and jeopardize the future of manned space exploration. Thus, NASA officials did in fact do what PR professionals are taught to do: they responded quickly to the incident, holding a press conference three hours after the explosion; they provided the media and Congress with accurate information; and they immediately appointed an investigation panel composed of members independent of Apollo 13. There were still members of NASA on the board because the highly technical nature of the mission meant that only its personnel and some of its contractors were privy to the mechanics of the spacecraft and could attest to what might have malfunctioned.

However, an additional difference between the Apollo 1 and Apollo 13 missions existed: the public availability of communication between ground controllers and astronauts. With Apollo 1, Jack King[105] had made a tape of the communications between the astronauts in the capsule and those in the blockhouse monitoring the preflight test. These tapes, though not released immediately by NASA, later revealed the three astronauts did not die immediately from the fire while sitting in their seats, as the agency had originally reported, but rather 15 seconds after the fire started. They perished from asphyxiation while desperately trying to escape the module.[106] It was understandable that NASA would not want to release this tape with the cries and pleas of the astronauts in their final seconds out of respect for the memory of the crew and for their families. However, NASA should not have concealed the fact that it had the tape, a communication only King knew had been recorded.

As a result, communication policies changed after the Apollo 1 tragedy to provide transparency. Later that year, NASA personnel at Kennedy Space Center introduced a "black box" system and made it available to the press through a leased line from the local telephone company.[107] Despite the fact that NASA project managers resisted this decision, it was nonetheless developed and installed. However, communications between astronauts and medical personnel were restricted to protect the privacy of the astronauts.[108] Black boxes were also installed in astronauts' homes so that the wives of the astronauts could listen to their husbands if they wished. In addition, after the Apollo 1 fire, NASA also allowed one print and one television reporter into Mission Control to cover special events.[109] These reporters, selected by the members of the press around the world, disseminated the information to pool reporters who further diffused the information to wire services; NASA's PAO only confirmed the accuracy of the reports.[110] This way, the press could make its own judgments pertaining to the space program.[111] In the end, NASA might have released the information and cooperated with authorities because it knew that the media had access to its communications anyway.

Thus, not only did NASA "come clean" with its handling of the Apollo 13 crisis, but it also actually strengthened its image in the eyes of Congress and the public. James Kauffman has argued that NASA, with the support of President Nixon, saved the day by trotting out metaphors of choice and by calling the Apollo 13 incident a "successful failure."[112] Once the Apollo 13 astronauts safely splashed down, for instance, NASA Administrator Thomas Paine said in a press conference that the astronauts illustrated "the great spirit of exploration, of bravery, men who are willing to venture out into new areas and overcome problems as we push forward to explore the unknown," as this "epic struggle" ended in "final victory."[113] Nixon's rhetoric mirrored Paine's, praising humans when machines failed, "A triumph of the human spirit—of those special qualities of man himself we rely on when machines fail, and that we rely on also for those things that machines cannot do." He added, "the astronauts did not reach the Moon, but they reached the hearts of millions of people in America and in the world."[114]

Emphasizing the human aspect of technology and the importance of Mission Control's teamwork in the Apollo 13 near-disaster, NASA illustrated to Congress and the public the utility of humans in space. This countered arguments that sending humans into space was costly and unnecessary. The space agency wanted to keep the gates open for future manned missions. Without humans in space, the frontier myth would not work; machines sailing the oceans of space simply do not create the same romantic images as individuals venturing through frontiers. This notion of "successful failure" made the public believe that NASA, despite its mistakes, could overcome any problem, and despite its mishaps, would remain technologically superior. And there was an additional happy ending to the story: Congress voted to add an additional $268 million to the 1971 NASA budget—more than Nixon had requested—funds that would begin the shuttle program.[115]

In sum, it is evident that NASA used a variety of public relations tools to convey its mission to Congress and to the American public: access to NASA and its communications, press releases, and materials such as film, photographs, and technical models. It also used symbols to communicate its existence, including insignias and flags. And just as any reputable public relations agency, NASA also used adverse situations to sell its Apollo program. Having learned to be forthright after the lessons of the Apollo 1 fire, NASA's Affairs Office was able to divert attention from the technical missteps of Apollo and instead create an image of a can-do agency. The "successful failure" illustrated that a near tragedy could be a triumph, a necessary image indeed for an organization dependent on public support and Congressional funding.

But more than its technical can-do image, with the help of imagery provided by the mission patches and the symbolism of Old Glory, NASA forged metaphors of the frontier and American ideals of exploring and conquering new territories to sell Apollo in its early days. Thus, while its press releases informed the media and the Spacemobile program educated students and teachers about the technical aspects of the Apollo program, it was the symbolism to which the American people could relate. While this chapter examined how NASA conveyed information to the press and the public, chapter 4 provides a discussion of how CBS News further presented NASA and its Apollo mission to the world.

NOTES

1. Gordon L. Harris, *$elling Uncle Sam* (Hicksville, NY: Exposition Press, Inc., 1976), 7.

2. James Clayton, "Apollo 11: Public Relations Dream," *Washington Post*, July 27, 1969, 38.

3. The History Channel, "Tech Effect: Apollo 11," originally aired June 20, 2004.

4. Stuart Ewen, *PR!: A Social History of Spin* (New York: Basic Books, 1996).

5. Jennifer Rudeseal Carter, "View from the Bird Watch: Media, Memory, and America's Mercury Astronauts" (PhD Dissertation, University of Southern Mississippi, Hattiesburg, MS, 1996).

6. Memorandum, Walter T. Bonney to the NASA Administrator, September 9, 1958, NASA Headquarters History Office, Washington, DC.

7. Howard Benedict, "Full-Court Press: Apollo Meets the Media," *Air & Space*, June/July 1989, 83.

8. Robert J. Donovan and Ray Scherer, "Live, from the Cape," *Air & Space*, October/November 1992, 68.

9. James Kitchell, NBC News reporter during the early years of the space program, quoted in "Remembering the Space Race: A Panel Discussion" held at the National Air and Space Museum of the Smithsonian Institution in Washington, DC on November 15, 2000. Transcribed in *Quest, The History of Spaceflight Quarterly* 9, no. 3 (2002): 6.

10. Ibid., 7.

11. Ibid.

12. Donnovan and Scherer, "Live from the Cape;" and *Quest*, 8.

13. NASA, "Functions of the Administration," *Space Act of 1958*, accessed May 1, 2024. https://www.nasa.gov/history/national-aeronautics-and-space-act-of-1958-unamended/.

14. Walter T. Bonney Memorandum to NASA Administrator, August 20, 1959, NASA Headquarters History Office, Washington, DC.

15. Ibid.

16. Ibid.

17. Walter T. Bonney Memorandum to NASA Administrator, September 9, 1958, NASA Headquarters History Office, Washington, DC.

18. Ibid., 8.

19. PIO and PAO are the same office. In the early days of NASA, the Public Affairs Office was referred to as the Public Information Office.

20. Walter T. Bonney Memorandum, August 20, 1959. It was also NASA's policy not to monitor interviews between NASA employees and the media, as stated in a memorandum from Scheer to the Field Center Public Affairs Officers, dated July 27, 1964. This can be contrasted with a memo to Bonney from Glennan indicating, "There is a need to exercise control over the public statements made by NASA staff," dated January 22, 1959. This was not to silence employees, but rather to control,

centralize, and unify statements by requesting employees refer questions to Public Information Officers. Employees were asked to refer requests for information to the PAOs in a memo from Scheer dated July 24, 1968, to field centers.

21. Walter Cronkite, quoted in Quest, 8. It should be noted as well that in a memorandum to NASA Headquarters Staff from NASA Administrator T. Keith Glennan, dated January 29, 1959, staff members were asked to "refer all inquiries from news media to the Office of Public Information" to coordinate the dissemination of information. NASA History Office, Washington, DC.

22. "Remembering the Space Race: A Panel Discussion," *Quest*, 9.

23. Paul Haney, *Newsweek*, July 28, 1969, 27.

24. Donovan and Scherer, "Live from the Cape," 70.

25. Walter Cronkite, "We are the Children of the Space Age: A Noted News Commentator Looks Back to Jules Verne and Forward to the Future," *TV Guide*, July 19, 1969, 10.

26. "Remembering the Space Race: A Panel Discussion," *Quest*, 12–13.

27. Edward Bliss, *Now the News: The Story of Broadcast Journalism* (New York: Columbia University Press, 1991), 367.

28. "Remembering the Space Race: A Panel Discussion," *Quest*, 14.

29. "Space History," Space.com, 2004, accessed January 9, 2004, http://www.space.com/news/spacehistory/scheer_obit_040901.html.

30. Donovan and Scherer, "Live from the Cape," 73.

31. Dick Dubrow, "Well, You've Seen it: Buck Rogers is Real," *The Evening Star*, July 21, 1969, D-5.

32. Don Hewitt, *Tell Me a Story: Fifty Years and 60 Minutes in Television* (New York: Perseus Books Group, 2001), 72–73.

33. Benedict, "Full-Court Press," 82.

34. Donovan and Scherer, "Live from the Cape," 68.

35. Ibid., 89.

36. Ibid., 87.

37. Cronkite, "We are the Children of the Space Age," 12.

38. Anonymous NASA PAO, Telephone Interview with the author, April 15, 2004.

39. One such request that Scheer thought was ridiculous was a network asking to land its helicopter on a military aircraft carrier to film a splashdown. Donovan and Scherer, "Live, from the Cape," 70–71.

40. "NASA's Public Information Policy," *The Washington Post*, March 27, 1971, A18.

41. Hugh Harris, telephone interview with author, April 2, 2004.

42. Letter to the Honorable Goodell from Robert F. Allnutt, Assistant Administrator for Legislative Affairs, NASA Headquarters, Washington, DC, January 16, 1968.

43. Harris, *$elling Uncle Sam*, 21–22.

44. Joel Banow, CBS News Director, in an interview with Martin Collins, Curator, Smithsonian Institution, National Air and Space Museum, Division of Space

History, 17. February 1, 2000. National Air and Space Museum Archives, Accession number 2000-0027, Garber Archive Facility.

45. NASA Press Kit, Apollo 8, December 15, 1968, Release No: 68–208, NASA Headquarters, Washington, DC.

46. The case was dismissed by the U.S. Supreme Court, which found that the astronauts' freedom of speech was protected by the First Amendment even while "on the job" in the U.S. government.

47. NASA Press Kit, Apollo 11, July 6, 1969, Release No: 69-83K, NASA Headquarters, Washington, DC, 1.

48. Ibid., 3.

49. Andrew Chaikin, *A Man on the Moon: The Voyages of the Apollo Astronauts* (New York: Penguin Books, 1994), 137–138, 618.

50. Ibid.

51. Charles Murray and Catherine Bly Cox, *Apollo: The Race to the Moon* (New York: Simon & Schuster, 1989).

52. Transcript of interview conducted by Rick Collin, NASA Aerospace Education Services Program Videoconference Coordinating Producer, with his father, Everett E. Collin, a former NASA Deputy Director of Education. The elder Collin managed the first years of the Spacemobile program. Session Three, August 6, 1992, 31. NASA History Office Files, Record Number: 12664, Washington, DC.

53. Ibid., 35.

54. NASA Facts: "Spacemobile," August 22, 1991. NASA History Office, NASA Headquarters, Washington, DC. File Record Number: 12664.

55. Hugh Harris, Telephone Interview with author, April 2, 2004.

56. Mark E. Byrnes, *Politics and Space: Image Making by NASA* (Westport, CT: Praeger Publishers, 1994).

57. "History of the Nike Swoosh and the Birth of a Multi-Billion Dollar Brand," MSN.com, The Street, accessed June 27, 2024, https://www.msn.com/en-us/money/companies/history-of-the-nike-swoosh-and-the-birth-of-a-multi-billion-dollar-brand/vi-BB1nl7Py?item=themed_featuredapps_enabled?loadin=defaultbrowser#details.

58. Whitey Smith, who was Executive Director of the Flag Research Center, quoted in Helen Cruikshank Sadtler, "Flags on the Moon," 98. On file at NASA Headquarters, Washington, D.C.

59. Howard E. McCurdy, *Space and the American Imagination* (Washington, DC: Smithsonian Institution Press, 1997), 90.

60. John F. Kennedy Library and Museum, 2004, accessed December 5, 2014, http://www.cs.umb.edu/jfklibrary/j091262.htm.

61. "The Outer Space Treaty of 1967," National Aeronautics and Space Administration (NASA), accessed June 18, 2024, https://www.nasa.gov/history/SP-4225/documentation/cooperation/treaty.htm.

62. "Richard Nixon Inaugural Address," The American Presidency Project, University of California, Santa Barbara, accessed June 18, 2024. https://www.presidency.ucsb.edu/documents/inaugural-address-1.

63. Thomas O. Paine, Memorandum to Associate and Assistant Administrators and Center Directors, February 25, 1969, NASA History Office, NASA Headquarters, Washington, DC.

64. Ibid.

65. Ibid.

66. Helen Cruikshank Sadtler, "Flags on the Moon," on file at NASA Headquarters, Washington, DC, 100.

67. "'Put an American Flag on the Moon—Bennett'," *Space Daily*, June 3, 1969, 148.

68. Anne M. Platoff, "Where No Flag Has Gone Before: Political and Technical Aspects of Placing a Flag on the Moon," *NASA Contractor Report 188251*, on file at NASA History Office, Washington, DC, 6.

69. Ibid.

70. Dorothy McCurdle, "The U.S. Flag from the Moon Voyage Furled at $25,000," *Washington Post*, June 14, 1970, 161. The winner was the wife of a Las Vegas, NV, publisher.

71. Sadtler, "Flags on the Moon," 105.

72. Ibid.

73. David Bowman, *This Must be the Place: The Adventures of Talking Heads in the Twentieth Century* (New York: HarperCollins Publishers, 2001), 220.

74. Instead, there was a voice-over with the narrator announcing, "Ladies and gentlemen, let's rock and roll."

"The 1980s: All Eyes Focus on Space Shuttle," NASA.gov, accessed June 30, 2024, https://www.nasa.gov/history/the-1980s-all-eyes-focus-on-space-shuttle/.

75. Judith Kaplan and Robert Muniz, *Space Patches: From Mercury to the Space Shuttle* (New York: Sterling Publishing, 1986), 35.

76. Ibid., 34.

77. Ibid.

78. Ibid., 48.

79. Ibid., 55.

80. Richard L. Lattimer, *All We Did Was Fly to the Moon* (Gainesville, FL: Whispering Eagle Press, 1985), 93.

81. Ibid., 94.

82. Ibid., 61.

83. Ibid., 62.

84. Kaplan and Muniz, *Space Patches*, 55.

85. Lattimer, *All We Did*, 77.

86. Lattimer, *All We Did*, 90.

87. Ibid., 94.

88. Byrnes, *Politics and Space*, 2.

89. Ibid., 7.

90. Ibid.

91. Ibid.

92. Ibid.

93. Ibid.

94. Ibid.
95. Ibid.
96. Ibid.
97. Ibid.
98. Ibid.
99. James Kauffman, "A Successful Failure: NASA's Crisis Communications Regarding Apollo 13," *Public Relations Review* 27 (Winter 2001): 437–438. https://doi.org/10.1016/S0363-8111(01)00099-6.
100. Ibid., 422.
101. Ibid., 424.
102. Ibid.
103. Ibid., 425.
104. The 1995 *Apollo 13* film misquoted the phrase by saying, "Houston, we have a problem."
105. Jack King was a Public Affairs Officer at Kennedy Space Center in Florida. He was also known as "The Voice of Apollo" because he announced the countdown time to liftoff for the Apollo missions.
106. Kauffman, "A Successful Failure," 425.
107. Gordon L. Harris, *$elling Uncle Sam,* 206.
108. Author interview with NASA Public Affairs Officer, who asked to remain anonymous, April 15, 2004.
109. Kauffman, "A Successful Failure," 439.
110. Ibid.
111. Ibid., 440.
112. Ibid., 441–442.
113. Ibid., 442.
114. Ibid.
115. Ibid., 445.

Chapter 4

Live from the Project Apollo and CBS's "Space Headquarters"

It is not surprising that pictured on the front sleeve of Steven D. Stark's *Glued to the Set: The 60 Television Shows and Events that Made Us Who We Are Today*[1] is an old floor model television with the now-famous image of Astronaut Buzz Aldrin standing in front of the American flag that had just been planted on the Moon, for the space program in many ways "grew up" with television.[2] Projects Mercury and Gemini, as pilot tests for the Apollo program, were the first televised manned space missions and were telecasted in black and white, while the last Apollo mission ended in color. Network producers filmed Project Mercury in 1961 using 16 mm film and advanced to videotape technology by the time the Apollo program ended in 1972. Coverage of Apollo began with mechanical animations of model spacecraft and ended with live transmissions from the Moon. Television progressed with the space program and was responsible for the aforementioned advances because the networks had to invent space-related animated broadcast technology as they went along to realistically cover the Moon shots. And what better match for television than the space program? The early days of the space effort provided everything essential for "good TV": drama, novelty, suspense, heroes, mystery, and humor. These combined aspects offered an escape from life's everyday routines and did so for hours at a time. CBS broadcast the Apollo 11 touchdown on the Moon for 34 consecutive hours, NBC for 31 hours, while ABC broadcast it for 30.[3] CBS, now owned by Paramount, held the number one rating for Apollo coverage, and still today, with 228 affiliates, the CBS Television Network is currently rated the most-watched television network, with a national reach of 96 percent of all U.S. households.[4] Its video-on-demand and streaming services enable viewers to watch programming at their leisure.

Altogether, an estimated 600 million people, or about one-fifth of the total world population in 1969, watched the Moon landing on television.[5] Clearly, most Americans learned of and followed the Apollo program on television, and the medium seemed to have met the task. But the networks had a lot at stake. Each of the big three invested over $1.5 million to broadcast the Apollo 11 mission alone.[6] It is no secret that news producers and reporters were not scientifically trained and therefore depended on other sources of information to meet their obligation to the public. As composer John Cage said, "The Moon landing will expand the media just as it is expanding our minds—that is, our sense of what we are capable of doing."[7] And expand it did!

Television fulfilled Schudson's promise that cultural memory will be created and maintained by cultural institutions. In this case, television networks did just that; they created and maintained narratives, which, of course, involves many factors that shape content and in turn dictate structures. In *Mediating the Message*, Pamela Shoemaker and Stephen Reese provide a model of media content as a guide that this author uses to explain how CBS News reported the Apollo story. Shoemaker and Reese argue that most media research has focused on the audience and effects of media content instead of the factors that shape media content. A wide variety of factors, both internal to and external to media organizations, are just as important as the effects on the audience: individual (a media worker's personal characteristics), media routines (patterns that enable efficiency in reporting), organizational (internal structures that influence content), external (organizations and factors beyond the control of the media organization), and ideology (socio-cultural concepts)[8] influence media content.

FACTOR 1: INDIVIDUAL MEDIA WORKERS

Individuals do influence the reporting of news.[9] For the task at hand, each network filtered information on Apollo through its lead anchor. In CBS's case, that was Walter Cronkite, whose face and voice soon became associated with the Moon shots. Cronkite was a space enthusiast from the beginning of the televised space programs. Here, the Shoemaker and Reese model for individual factors is evident in Cronkite's personal beliefs, attitudes, and values as he was an enormous fan of and strongly believed in space exploration. In addition, as Cronkite voiced on several occasions, his professional background, particularly in covering World War II, enabled him to cultivate an interest in aeronautics. Born in Missouri in 1916, Cronkite was educated at the University of Texas and began his journalism career at the *Houston Post* before moving to the United Press (later known as UPI) in 1937. He covered World War II, parachuting with the 101st Airborne Division, flew a bombing mission over Germany, and later covered the Nuremberg war crime trials. Cronkite's fascination with rockets began when he watched the German-made V-2 rockets launch from Wassenaar in the Hague region of the South Holland province of the Netherlands. After the war, Cronkite set up United Press's first postwar Moscow bureau, where he was chief correspondent from 1946 to 1948. Switching to the entertainment side of the network in 1960, Cronkite joined CBS where he was the host of *You Are There*, a historical recreation series, and co-hosted the *CBS Morning Show* with puppet "Charlemagne." But it was really as an anchor for the 1952 presidential conventions that Cronkite became well known as a journalist.[10]

Alan Shepard's first space flight on May 5, 1961, allowed Cronkite the newsman to put his journalistic skills to use covering a topic that he regarded with great interest. Cronkite admits to getting "caught up" in the excitement after witnessing failure upon failure of the early American rockets blowing up on launch pads. Even though NASA provided assistance with press releases and film, "television coverage per se for the space program would in effect have to invent itself."[11] Cronkite recalled it was only himself, a producer, a cameraman, and snakes in the marshy Cape Canaveral who provided the CBS coverage in the beginning years of the space program.[12] But by the time the Gemini project was developed in 1965, the networks invested in more permanent facilities that would accommodate additional reporters and equipment. Cronkite's desk was situated in front of a large glass window that allowed television audiences an unobstructed view of the launch pad in the distance, which could be seen in CBS's coverage of the Apollo missions.

In April 1962, Cronkite became an anchor for the *CBS Evening News*, which was expanded from 15 to 30 minutes less than a year later. Because NBC's Huntley-Brinkley report was the ratings leader, CBS replaced Cronkite briefly for the 1964 presidential conventions, but public outcry protesting the change forced CBS to allow Cronkite to remain as anchor. In 1966, CBS began to take the lead from NBC and in 1967 became first in the ratings. CBS feared its audiences would turn back to Huntley-Brinkley for space coverage, but it was Cronkite's authoritative reporting on space activities that caused many viewers to follow the American space program through Cronkite. In fact, many believed Cronkite overtook Huntley-Brinkley specifically because of his space reporting, on which he worked "tirelessly."[13] To illustrate, knowing he needed information at his fingertips, especially during delays and last-minute cancellations, Cronkite collected data from NASA, its contractors, and federal agencies on the astronauts, the spacecraft, and the rockets. He also visited the manufacturing sites of space contractors and even trained as the astronauts did. He experienced Moon gravity at Langley Field (now called Langley Air Force Base) and total weightlessness at Patterson Air Force Base, landed a lunar lander mock-up, and drove the lunar rover, all of which he demonstrated for television audiences. Later, Cronkite stated that he participated in these activities because "they have a special fascination for me" and "because I have a desire to report on the sensation."[14] Cronkite even catalogued his experiences in a diary that was typed in large letters so he could use the information when he needed a quick reference while on air.[15]

Often called "the fourth astronaut" in the Apollo program,[16] Cronkite was the man Americans turned to him for space coverage because of his extensive research and expansive knowledge, which paid off for CBS. For the Apollo 11 mission, Arbitron ratings placed CBS in first place with 45 percent of the national audience, with NBC at 34 percent, and ABC at 16 percent.[17] So popular was the anchor that when John Glenn's mother visited the Cape, she said above all she wanted to see Walter Cronkite.[18] Often referred to as "the most trusted man in America," Cronkite ended his broadcasts with ". . . and that's the way it is . . . ," and people believed him. Stoic on the surface, Cronkite admitted, "I am bland and deliberately so. Perhaps in all the Nation's turmoil and trouble, the viewers want blandness."[19] Because of his grandfatherly image and plainspoken news reporting, the public identified with him more than any of the anchors. As columnist Art Buchwald asserted,

> My wife identifies with Mr. Cronkite even more than she does with the astronauts. If Walter's relaxed, she's relaxed—if Walter gets nervous, my wife gets nervous. Walter Cronkite, more than anyone else, sees us through these shots and we really count on him to get the Apollo capsules back safely to earth.[20]

Viewers even wrote to NASA asking the agency to officially name Cronkite an "honorary astronaut."[21] Others in the industry also recognized Cronkite for his dedication. For the Apollo 11 broadcast, 52-year-old Cronkite was on air for 31 of the 34 hours of continuous CBS broadcast of the event, a feat that earned him the nickname "Old Ironpants."[22] Cronkite said he was "so interested and excited that I don't get tired—I get too emotionally involved."[23] And he did not hide his enthusiasm for the space program, which he thought demonstrated "that particular combination which represents America at this best: pragmatism with the dogmatic objective, hard-headedness with a soft heart."[24]

Although CBS News Headquarters in New York had the final say in production matters, people who worked closely with Cronkite remember times when he ignored modifications to a broadcast while reporting live on air, and the New York offices went along with Cronkite in his decision.[25] This was because Cronkite, with his superior knowledge, was thought to be an expert on space matters even to CBS's top decision-makers. Even so, Cronkite was criticized for being a cheerleader for NASA:

> I was accused of having failed to observe the first journalist precept of impartiality, of having unabashedly been a NASA booster. I believe my critics were wrong. I know that I did not try to suppress my excitement about the technical achievements and the grand adventure of space flight, but I also disagreed sharply with some aspects of the space program. NASA didn't bestow any medals on me.[26]

Well, not at that time, anyway. Even though he failed first-year physics at the University of Texas,[27] Walter Cronkite later became one of the most recognized and respected "science reporters" in the history of broadcast news. In fact, while the Apollo program was drawing to a close, Cronkite received a Museum of Science award for "outstanding contribution toward public understanding of science."[28] Inducted into the Academy of Television Arts and Sciences Hall of Fame in 1985, Cronkite was also awarded the Corona Award from the Rotary National Award for Space Achievement Foundation in 1999 for "a distinguished lifetime of achievement in the field of space" after two decades of space reporting.[29] The *only* manned mission Cronkite did not cover was a Skylab mission that flew while he was on vacation.[30] So respected is he that Cronkite is the only journalist to have been voted among the top ten "most influential decision-makers in America" by *U.S. News and World Report*, and has been named "the most influential person in broadcasting."[31]

Not only did Cronkite's space coverage win two Emmys, but it also enabled him to become a semi-finalist in NASA's Journalist in Space[32] program before it was canceled after the January 28, 1986 Challenger explosion that killed all seven aboard, including Christa McAuliffe, winner of the Teacher in Space program. In 2004, Cronkite was finally given a NASA award, along with NASA's first generation of astronauts, and was named an "Ambassador of Exploration" whose recipients "help communicate the benefits and excitement of space exploration and why the continuing investment in our future is vital to the security and vitality of America."[33] The award was presented at a ceremony celebrating the 35th anniversary of the Moon landing, and Cronkite was the only non-astronaut to receive it. The space agency also asked Cronkite in 1993 to narrate its 20-minute video production "Strategic Planning: Charting a Course for the Future" to explain and promote strategic planning to government managers and employees because of his "comforting and authoritative voice."[34]

Even today, several decades after the Apollo program ended, it is apparent both in contemporary media representations and in individuals' memories of the Moon landing that the Apollo story cannot be told without reference to CBS and Walter Cronkite. And it is, of course, also evident in sites whose purpose is to remember Apollo. For example, in 2003, George Mason

University's Center for History and New Media invited Internet users to post their most vivid memories of Apollo 11 for the mission's upcoming 25th anniversary. Overwhelmingly, respondents described in detail the exact location of the television set in their homes, and more importantly, what they saw on its screen: Cronkite and his reactions. "Tears came to his eyes, and I cried as well." "We sat about two feet from the television set completely enthralled by Walter Cronkite's every word." "Feeling like I was suddenly thrust into the 21st Century with Walter Cronkite." "My most vivid memory was Walter Cronkite crying. The only other time I can remember this happening was when Kennedy was killed."[35] Cronkite and CBS were part of the Apollo experience, writ large.

FACTOR 2: MEDIA ROUTINES

The next element of Shoemaker and Reese's model addresses media routines and how such routines influence content. Routines are repeated patterns that enable media workers to do their jobs efficiently, especially in television, which has a high turnover rate. Routines become especially important here because much of the equipment used is highly specialized, including satellite transponders, computerized editors, and digital translators, to name a few. The result of knowing how to use these kinds of equipment is a "technically uniform, visually sophisticated, easy-to-understand, fast-paced, people-oriented" story that is produced in a minimum amount of time.[36] Other constraints that influence media routines are what the media outlet is capable of producing, what media products audiences are willing to accept, and what sources of information are available for the media outlet to use.[37] What audiences were willing to accept was not an issue here because public demand for information was such in the early years of Apollo that the networks were hard-pressed to supply enough.

"Live" coverage was new, so routines became even more important for networks to put together a story by deadline. Because live shots in the early 1960s were very expensive, the three networks, ABC, CBS, and NBC, pooled resources to share financial burdens and the limited space available to reporters covering the launches. Pooling resources also enabled the networks to devote more money to special effects coverage. The networks collectively decided what they needed from NASA to produce a news segment, such as whether NASA would provide a microphone for the liftoff sound, for example.[38] Thus, networks working together made routines easier to manage. Video feed that NASA supplied to the networks, such as liftoff footage, was the same. What differed were the three individual anchors' descriptions of the events of the day.

Each of the networks also tried to differentiate itself in terms of comprehensive coverage, hoping breadth would earn the best ratings. CBS strove for a unique look through the use of opening titles, sometimes called opening storyboards, which were images projected against the backdrop of the news anchor. In such a shot, the anchor is in the foreground and the opening title is in the background. Opening titles alerted CBS viewers visually. Although its competitors also used opening titles, Joel Banow, former CBS Special Events Director, believes CBS's storyboards were more creative, more realistic, and more visually appealing. In fact, Banow credits the storyboards with CBS's high ratings for Apollo coverage.

> All of the networks were essentially the same; we all had the same access to the astronauts, etc. CBS and others created a bank with information and would call up the bank . . . to get the materials they needed for a certain broadcast. So, news-wise we were not that different. The difference was in how we portrayed the mission visually, how we presented it. It was my intent to make it as realistic as possible and as scientific as possible. The other networks had to use the same materials—same models, etc.—but it was the sophistication we had at CBS, with the talent of Dick Spies who created 35 mm cell animation. We made sure the sun was where it should be, for example, and everything three-dimensional, and it was airbrushed animation to make everything look real. I treated each event as if it were a movie. I wanted to combine all the sources of images to create a montage of different points of view. That's where we differed. We had much more depth—depth of our reporters in terms of going all over the world. We picked the right people to contribute their input, giving us more depth and sophistication to our coverage.[39]

To understand the Apollo missions, one must go back to a time before VCRs, before computer-aided design models, and before portable video cameras. Because NASA provided well in advance a minute-by-minute account of each Apollo mission, it is difficult to imagine any event in the history of television that was as scripted as the Apollo missions. Yet coverage needed to be thoroughly and exactly planned because of its technological complexity. Indeed, CBS came a long way from the early days of space flight when Cronkite and cameramen operated from the back of the network's station wagon.[40] Realizing it needed a department that could produce shots in advance, the CBS special events unit was created shortly after Sputnik launched (the unit also worked on campaigns and elections). This department would use NASA flight information to plan broadcasts accordingly—factors such as the importance of the mission, required guests to explain scientific material, essential background information of the mission, and simulations of the flight plan, for examples.[41] To illustrate, Banow described how a typical Apollo mission was planned:

> We sit down, we analyze the mission and what we need. Okay, well, I've got prior stuff from missions that I can reuse. I probably have an Apollo launch animation. That's no problem. Of course, I don't need it initially, but once it goes out of sight, I've got to cut to something, so I have the animation I can go to at a higher-altitude stage. . . . We've never had translunar injection . . . so now I need a new animation. I've got to show that burn . . . so I've got to research that burn, how long is the burn, and I . . . make an animation that's exactly the right burn time . . . I see what I need and I make my requests. . . . I would usually create a memo that this is what I need for this mission and pass it on to my boss and he usually approved it. . . . And then it would be my boss' job to go and sell the network or CBS News execs and then the network and saying, "We need twenty hours of programming." . . . We always got the airtime, and we got the budgets and that was it.[42]

To further illustrate its "CBS News Simulations," the network needed a model of the Moon, again, not an easy task. For $11,000 (1969 dollars), CBS contracted with Educational Frontiers to create a lunar moonscape on which the network could place the lunar lander model it had built to explain complex movements in space.[43] The lunar model looked nearly identical to the Moon's surface in photographs, such as the one taken during the Apollo 8 mission in December 1968 (figure 4.1). CBS also acquired a continuously moving seven-foot-long rubberized conveyor belt to simulate orbiting the lunar surface[44] (figure 4.2), and its surface also very closely resembled that of the Moon. CBS was the only network to own such a contraption.

In a pre-computer-aided animation era, visualizing space travel over a quarter of a million miles in a scientifically accurate way was difficult.[45] Here, television producers had to find a way to demonstrate an engine burn given the fact that there is no oxygen in space.[46] Producers, then, needed to develop ways to illustrate unusual laws of physics, such as depicting how a spacecraft slows down in order to speed up in space.[47] A Bunsen burner was used to mimic rocket engines and the spacecraft's heat shield upon re-entry into the Earth's atmosphere.[48] Broadcasts for each mission tried to improve on the techniques employed to depict each previous mission, to make each mission look even "more real" and "more exciting."[49] On the other hand, using the flight plan from NASA as the script, CBS had months, sometimes up to a year,[50] to prepare for its broadcast. Cronkite remembers the Apollo 11 Moon landing coverage:

> It was timed, of course, precisely to the flight plan, so that we rolled this prepared tape, and it was to the split second of what was happening up there on the Moon. And, of course, we were getting reports and the audio report from the landing as to how it was along in little bits and pieces. If there was a little space, I would try to fill in what the situation was at the moment.[51]

Figure 4.1 Lunar Surface Photo from Apollo 8 Target of Opportunity. ***Source*****: Courtesy of NASA.**

Even so, Cronkite remembers being speechless when the lunar module landed—the result of tension that occurred when it appeared that it might not land. Armstrong and Aldrin were headed for a large boulder and narrowly missed landing on it. They almost ran out of fuel and came within 16 seconds of the mission being aborted. Routines were important because they enabled media workers to cope with the unpredictability of live broadcast television. Banow, who in 1969 received a Directors Guild of America award for his coverage of Apollo 11, explained how CBS contended with the landing:

> We're sitting there, and we're still hearing, you know, 20 feet down to the left, two to the right, and then, oh, we're down already. But, through the luxury of videotape, and the fact that we had to go back on the air later and recap the landing on the Moon, we were able to go in and sync up the audio. When we had them land [in the recap], they landed exactly on the right time, with exactly the right word from that point on throughout our coverage.[52]

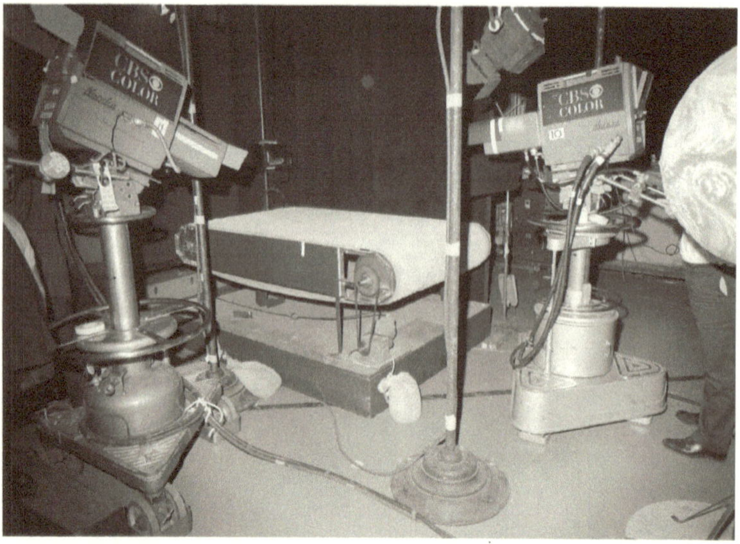

Figure 4.2 Lunar Conveyor Belt, Apollo 11 Coverage. ***Source:*** **CBS/Getty Images. Used with Permission.**

Should routines fail, networks need a contingency plan because in live television broadcasts, there can be no downtime. Having a backup plan, then, also becomes routine in televising space exploration.

FACTOR 3: ORGANIZATIONAL FACTORS

Organizational factors also influence media content. Here, Shoemaker and Reese describe the internal structures of a media outlet and show how such structures impact the organization's goals, markets, and technology. For instance, how many brands are owned by the parent organization, and how might this ownership factor into advertising content? In addition, what are the hiring and firing practices of an organization? Is the organization locally owned, or is it a chain that dictates content? CBS's primary goal in the 1960s was to take away the number one slot from NBC, a rating earned by the Huntley-Brinkley Report's coverage of the 1960 primaries. CBS was determined to be first in coverage of the space race and extensively covered all space stories, beginning with the Gemini program in the early 1960s. The popular Cronkite stayed on air for days at a time during space launches as part of that strategy, as described above.

For the Apollo 11 coverage, CBS clearly had the largest percentage of the audience. However, after the Apollo 11 Moonwalk, Nielsen and Arbitron ratings and CBS in-house research showed that audiences began to dwindle as the public began to lose interest in the space program.[53] After all, the mission had been accomplished; we had made it to the Moon before the decade was out. Thereafter, although the networks continued to cover the actual launches, they did so without the massive background information, historical data, and global perspectives as they had for Apollo 11. The two exceptions were the Apollo 13 explosion in space and the Apollo 15 mission with its $40,000,000 lunar rover, known as the 10 mph "Moon buggy," that captured the public's attention. Aside from those two missions, after the climax of Apollo 11, CBS planned its programming according to the interest of the audience, or in this case, the lack thereof.[54]

FACTOR 4: EXTERNAL FACTORS

The fourth element in Shoemaker and Reese's model includes factors external to the media organization that also influence content. First are external sources of information, such as PR campaigns and special interest groups. Shoemaker and Reese indicate that official sources are usually the dominant ones; an example is an "expert" on a topic. Government and businesses can influence content by offering the media outlet information that otherwise is unobtainable. Technology also affects content, such as providing better graphics and photographs. Access to specialized databases can offer a media worker more information to enrich his or her stories. Finally, the economic environment influences content as well as revenue. For example, if there is more than one newspaper in a town, a company's advertising dollars must be split between the two newspapers, thereby limiting content. Advertising, however, was not an issue for the space shots. Although it might be assumed that the networks lost money covering the Apollo missions because of the interruption of scheduled programs, advertisements were scheduled during the breaks in coverage. CBS News did not lose money.[55]

Because most broadcasters were not trained in the sciences, reporters depended on external sources for the technical information needed to tell the Apollo story. Network employees did have close working relationships with NASA and its public affairs department because the goals of both NASA and the networks were to inform the public about the space program. Banow describes the familial relationship between NASA and CBS:

> There were a lot of people [who] became very friendly, as all the press people did, no matter who you were, producers, researchers, correspondents, anchors, in that there was a very close relationship with the NASA public relations office and its personnel, whether at the Cape or whether in Houston. . . . Down at the

> Cape in those days, being at Fat Boy's, munching on ribs, everybody, it was one big crazy family in those early days.[56]

Much of the information that the networks received came from NASA, including pre-packaged video clips. Although reporters such as Walter Cronkite added interpretations to the video, the footage was obviously what NASA wanted the public to see. Moreover, CBS hired Cronkite's close friend, Mercury/Gemini/Apollo astronaut Walter "Wally" Schirra, as a guest space analyst to "interpret" instructions "between the spacecraft and Mission Control" that Cronkite "didn't understand."[57] Schirra was needed, according to Cronkite, because he provided "a subtle and gently firm way of bringing broadcasters back to Earth whenever they tend to become over-dramatic or boyishly breathless."[58] Others, however, claim that Schirra, briefed by NASA before appearing on CBS with Cronkite, was well equipped with facts and figures that would offset any criticism of NASA, so much so that he seemed "remarkably prepared to deal with a wide range of touchy issues, from the costs of the National Aeronautics and Space Administration relative to the cost of other national programs, to the role of women in the over-all space program."[59] Cronkite, however, would remind critics that Schirra's pro-NASA bias "when kept within legitimate bounds, [was] to be expected."[60]

Technology was also crucial. Many television stations did not have advanced video and graphics systems. To compensate, NASA footage supplied to the television stations simulated what was happening in space. Businesses funded by NASA also contributed; contractors such as McDonnell Douglas and Rockwell supplied video and text describing the complex processes of putting men in space and getting them to the Moon and back successfully. NASA furnished detailed diagrams of spacecraft that enabled producers to create special effects.[61] Better yet, once the spacecraft were equipped with cameras in later Apollo missions, the producers could alternate between live pictures from the Moon and simulations using models.[62]

The opening titles of each of CBS's newscasts were also modeled after NASA-inspired "PR" tools, such as the patches described earlier. CBS was the only network that produced full-cell animated titles of the patches made to look three-dimensional by artist Richard Spies,[63] "who understood the scientific principles of space flight and astronomy."[64] Producer Joel Banow created the titles using the patches because each of the missions' purposes was unique.[65] These patches, incorporating NASA ideology, were highly symbolic. A patch "branded" the mission and the newscast. For example, for the Apollo 14 mission, whose command module was named "Kitty Hawk," Banow designed a cell featuring the original Wright Brothers plane with a live shot of the Moon superimposed over the patch for the broadcast.[66] This visual graphic of the aviation-first capitalized on the United States' tradition of technological firsts. At other times, the opening title featured a photograph of the three astronauts for each of the Apollo missions. None of the three crewmembers would fly a second Apollo mission together, and few of the Apollo astronauts flew a second Apollo mission. Thus, like the patches, the faces and names of the astronauts "branded" each mission.

Although NASA and its contractors who built the Apollo spacecraft provided the bulk of information to explain how the spacecraft worked, on occasion CBS arranged to create a special simulation.[67] NASA and its contractors also provided models of spacecraft, but when it needed additional models, CBS built them in-house based on specifications supplied by NASA and/or the contractors. CBS hauled in tons of dirt for a simulation of the lunar rover after Grumman supplied the network with a remote-controlled model of the Moon buggy.[68] Pictured (figure 4.3) is the Lunar Roving Vehicle on the Moon during the Apollo 15 mission.

During and after the launch, there was always direct communication between CBS and NASA. Telephones were installed at the network that signaled with a blinking light that new information was available pertaining to a launch. The light was called the "Haney Light," named after "The Voice

Figure 4.3 Lunar Roving Vehicle. ***Source:*** **Courtesy of NASA.**

of NASA" Paul Haney, public affairs director at the Houston installation that housed Mission Control. Telephones signaled to television producers that a Mission Control announcement was imminent so the information could then be passed on to the news anchor.[69] But the space agency and its contractors were not the only sources of information for CBS researchers. The network also contacted universities, museums, and observatories, and even hired technical advisors, scientists, and astronomers independent of NASA who could explain detailed scientific data to network personnel.[70] CBS also consulted books and newspaper articles pertaining to science and technology, as well as magazine features from *Business Week, Newsweek, Aviation Week, Space Technology,* and *Popular Science* to produce background information for the Apollo newscasts.[71]

FACTOR 5: IDEOLOGICAL INFLUENCES

The final component of the Shoemaker and Reese model is ideological influences, which they define as "a symbolic mechanism that serves as a cohesive and integrating force in society."[72] Shoemaker and Reese are interested in knowing the media's role in propagating that ideology. Ideology is important because it can reveal why a media organization chooses certain content when we cannot see the behind-the-scenes decisions made by editors, producers, and owners. In addition, ideology governs how the society sees itself, providing frames of reference to understand its culture. It articulates and reaffirms social values, working through existing values, and it works at a societal level, not an individual level.[73]

An example of how ideology influences media content that Shoemaker and Reese highlight is the coverage of the Vietnam War. The war was ideologically framed in several ways: "our boys to our rescue," "heroes in action," "war as an American tradition," "winning is everything," and "war is manly," to name a few. These ideological frames of reference, operating in the reporting of the Vietnam War, pushed the public toward consensus. Consensus, in turn, purged discourse of the moral and political implications of the war. For instance, news editors rejected David Halberstam's war reports and instead accepted more optimistic views of the war from the Pentagon and administration officials.[74] Thus, "not only is news about the powerful, but it structures stories so that events are interpreted from the perspective of powerful interests."[75]

Rhetorical analyses help uncover the story structures to get at dominant interpretations and powerful interests. Ideology shaped the rhetoric of space as much as that of the Vietnam War. The American people had to be convinced that it was crucial to win the Space Race, for if the Soviet Union controlled space, then the enemy would control the Earth. The ideological frames were not much different from those used in the Vietnam War coverage: "our boys to our rescue," "heroes in action," "America is number one," "winning is everything," "exploration (instead of war) as an American tradition," as will be seen in the next chapter. Using metaphors as guides in this rhetorical analysis will illustrate how these ideological frames functioned between 1968–1972, and how they compare to the ideological frames in contemporary representations today.

As with NASA, the goal of CBS, according to Joel Banow, was "to educate the people on our space exploration of the Moon. We wanted to try and talk about the Moon, its meaning, its place in history and so forth . . . as a public service [we] had to do."[76] As Banow continued to explain,

> [CBS] felt . . . that [Apollo] was kind of an exciting story and that we wanted to bring people up to date on what our government and what NASA was doing in lunar exploration. You know, there's still kind of a science fiction aura to those early years—talking about these things and doing them.[77]

Providing the public with news of hope during a time of tragic news (Vietnam, Chappaquiddick, assassinations, poverty, inflation, and race riots) may have been the network's way of giving audiences something for which to look forward.

Just as important, covering the Apollo program the way NASA wanted it covered allowed the networks to receive more information in return. Networks and NASA agreed that Apollo was exciting. Julian Scheer, head of NASA's public affairs, stated, "There was a collective excitement, I think, that ran through this great mass of population. Everybody felt a part of it; everybody felt they had a part. Everybody worked toward this goal. It was very unifying, [an] almost religious experience."[78] CBS played up the religious angle. According to a network representative, the space program highlighted "religion" as a shared belief in obtaining world peace, not just Scheer's "religion" of working toward a specific goal of landing man on the Moon:

> When we look at that Earth of ours out there like that, and I think they got a religion of their own when they saw it, and we should all, I think, get a certain religion to understand that we're on a fairly precious planet, a precious, precious planet with life on it, life with all that means.[79]

In other words, NASA viewed the accomplishment as a goal of the country and the agency; CBS saw it as a goal for the world.

As with its coverage of Vietnam troops, CBS portrayed the astronauts as all-American boys and was seldom critical of the space agency during the Apollo years. Instead, the focus was on the astronaut as a hero fighting the Cold War for God and country. As will be evident in the next chapter,

CBS focused on the astronauts living the American dream: most from the Midwest, the first in their families to graduate from college, married (with the exception of Harrison Schmitt) with two or three children, living in a nice house in the suburbs. CBS Executive Producer during Apollo, Robert Wussler, himself lived the "American dream." He began his career in the mailroom at CBS and moved up "through hard work and effort" to one of the network's exalted positions.[80] Former CBS producers, however, contend that CBS News did not make conscious efforts to portray the astronauts or anyone else in the space program as stellar American citizens to "give the audience what it wanted," as Banow explained.

> The idea that the astronauts were "the right stuff" began with the Mercury program and persisted through Apollo. They were test pilots and should have been there. The perception of the astronauts wasn't really manufactured; they boosted their own image just by being who they were. It was the media, especially the print media, who treated them like movie stars. They really were humble, and we tried to be as honest as possible. And I think we were as straightforward as possible.[81]

Given *Life* magazine's portrayal of the astronauts, it might be assumed, then, that NASA generated the boy-next-door image of the astronaut and that the media presented to the public what NASA gave it. However, "NASA knew they needed the networks, but they always gave us access to [the astronauts] and to their contractors,"[82] according to Banow. Just as NASA has claimed, CBS said that its top priority was:

> to get [the story] right. We made a conscious effort to do it right. But on the other hand was the entertainment side. This is where the special effects came in—the public wanted the Apollo story, and it was our job to present it as accurately as possible.[83]

NASA's ideology was evident in some of its PR tools, such as the aforementioned patches that CBS used as opening titles for an Apollo story. And it was through the patches that CBS far outshined its competitors, according to Banow, because the other networks did not have the graphic artists and the "imagination and vision" that Banow and former Walt Disney animator Richard Spies had.[84] For example, before the Apollo 11 astronauts ever stepped foot on the Moon, Banow and Spies designed an animation cell that featured an astronaut standing beside the LEM, with a reflection of a second astronaut in the gold faceplate of his helmet; "it turned out to look very much like the real thing. We weren't far off."[85] The patches, in addition to differentiating CBS from its competitors, also illustrated CBS's philosophy, according to

Banow. The patches were "wonderfully and graphically designed" because "we wanted something different for each mission, just like NASA did. What Dick Spies and I did was expand the patches—we utilized the symbols to reinforce the meanings of missions."[86] Banow, for instance, opened the Apollo 13 storyboard with Stonehenge followed by images of the Moon, and segued to the Apollo 13 patch that symbolized a chariot moving across the Moon. The animation then lifted the three stars representing the astronauts from the patch and placed them on the Moon.[87] This title improved upon the NASA patch by circumventing the mythology of the program in favor of a focus on the dreams of prehistoric cultures, tracing a route from Stonehenge to the Moon. CBS's symbolism illustrated the technological progression of man—American man.

The symbolism blossomed again with Apollo 17, the final Apollo mission. Replacing the patch, however, the Apollo 17 storyboard offered CBS's interpretation by foregrounding the legend of Icarus and his artificial wings. Here, CBS superimposed a balloon montage on images of early Wright Brothers flights and then infused images of Robert Goddard with his rockets into live images of current Apollo launches. These images morphed into pictures of men landing on the Moon and ended with images of Skylab, the project that followed the Apollo program. Rather than ending with the footage of Apollo 17, Banow decided instead "to look ahead to what was coming up in space and include that image of Skylab. . . . I think [that] just says something about the way CBS News special events thought."[88] This decision might, of course, have been a ploy to solicit interest for subsequent CBS broadcasts.

CBS stood to profit from the technological success of NASA. Proud of its coverage, CBS News released a 33 1/3 disk recording of the Apollo 11 mission entitled "Man on the Moon" narrated by Walter Cronkite, and an oversized book called *10:56:20 PM EDT 7/20/69: The Historic Conquest of the Moon as Reported to the American People by CBS News over the CBS Television Network* that featured dialogue and photographs of the Apollo 11 coverage. Like NASA, whose primary purpose was to inform the public, CBS also saw the importance of capitalizing on its success.

Now that a behind-the-scenes knowledge of how the *CBS Evening News* projected its content has been presented based on individuals, media routines, organizational structures, external factors, and ideology, part II offers an analysis of the final product manufactured by the network that reported from "Space Headquarters." In addition, the subsequent chapters offer an analysis of how the Apollo story has remained a topic of interest on CBS, CNN, PBS, and THC, and how the initial coverage has been refurbished in modern-day representations.

NOTES

1. Steven D. Stark, *Glued to the Set: The 60 Television Shows and Events that Made Us Who We Are Today* (New York: The Free Press, 1997).

2. Walter Cronkite ended his broadcasts with, "This Walter Cronkite, CBS News, Space Headquarters, New York" when presenting Apollo stories from 1969 through 1972. Other CBS anchors such as David Schoumacher did as well.

3. Edward Bliss, *Now the News: The Story of Broadcast Journalism* (New York: Columbia University Press, 1991), 369. See also Doug James, *Walter Cronkite: His Life and Times* (Brentwood, TN: JMPress, 1991), 125.

4. CBS, accessed February 21, 2024, http://www.cbs.com.

5. Harry Hurt, III, *For All Mankind* (New York: The Atlantic Monthly Press, 1988), xi.

6. "Great Adventure," *Newsweek*, July 21, 1969, 73.

7. Composer John Cage quoted in "Great Adventure," *Newsweek,* July 21, 1969, 75.

8. Pamela Shoemaker and Stephen Reese, *Mediating the Message: Theories of Influences on Mass Media Content*, 2nd ed. (White Plains, NY: Longman Publishers, 1991).

9. Ibid.

10. "Cronkite Remembers," Museum TV, accessed June 19, 2024, https://museumtv.pastperfectonline.com/archive/7438F209-966D-4BE2-A497-163975494915.

11. Doug James, *Walter Cronkite: His Life and Times* (Brentwood, TN: JM Press, 1991), 125.

12. Ibid.

13. Ibid., 126.

14. Ibid., 128.

15. Ibid.

16. In some cases, the "other astronaut" was also used. James, *Walter Cronkite,* 128. The "fourth" refers to the Apollo project as each of the Apollo missions had three astronauts. Cronkite has also been called "the eighth astronaut" as an addition to the original astronaut corps, in reference to the Mercury program. See Hurt, III, *For All Mankind*, 73.

17. Fred Ferretti, "Cronkite on Endurance: 'You Don't Think of That'," *New York Times,* July 24, 1969, 75.

18. Ibid., 73.

19. Lawrence Laurent, "Moon Mission Ratings Race," *Washington Post*, July 25, 1969, C7.

20. Art Buchwald, "When it Comes to Moon Shots, Just Put Your Faith in Walter," *Washington Post*, May 29, 1969, A19.

21. Thelma Jones's letter to Julian Scheer, NASA Chief PAO, March 18, 1969, NASA History Office.

22. James, *Walter Cronkite*, 132.

23. Ibid.

24. Walter Cronkite, "We are the Children of the Space Age: A Noted News Commentator Looks Back to Jules Verne and Forward to the Future," *TV Guide*, 1969, 10.
25. James, *Walter Cronkite,* 129.
26. Walter Cronkite, *A Reporter's Life* (New York: Alfred A. Knopf, Inc., 1996), 280.
27. Ibid.
28. "Cronkite Feted by Scientists," *Denver Post*, November 22, 1971, 29.
29. "Corona Award," Rotary National Award for Space Achievement, accessed June 19, 2024, https://www.rnasa.org/other.html.
30. James, *Walter Cronkite,* 128.
31. "Meet the People: Walter Cronkite," PBS.org, accessed June 19, 2024, https://www.pbs.org/avoidingarmageddon/meetThePeople/meet_03_03_01_cronkite.html.
32. A. R. Hogan, "Walter Leland Cronkite," *Ad Astra*, July/August 1991, 35.
33. NASA News Release 04-229, July 16, 2004. NASA History Office, 2.
34. "And that's the Way it is," *Government Executive*, January 1997, 4.
35. "Center for History and New Media," George Mason University, accessed February 4, 2004, http://chnm.edu/tools/surveys/details/70.
36. Shoemaker and Reese, *Mediating the Message.*
37. Ibid.
38. Joel Banow, oral history interview conducted by Martin Collins, Naples, FL, February 1, 2000, 33.

On file at the Smithsonian Institution, National Air and Space Museum Archives, Joel Banow Collection. (Hereafter cited as Banow-Collins oral history interview).

39. Joel Banow, telephone interview with author, March 8, 2005, and July 10, 2024.
40. "Remembering the Space Race," *Quest*, v.9, n. 3, 7.
41. Banow-Collins oral history interview, 17.
42. Ibid., 19–20.
43. Ibid., 34–35.
44. Ibid.
45. "Remembering the Space Race," 11.
46. Ibid.
47. Cronkite, *A Reporter's Life*, 280.
48. Artis Henderson, "To the Moon and Back," *Gulfshore Life*, accessed July 1, 2024, https://www.gulfshorelife.com/2019/07/30/to-the-moon-and-back/.
49. Banow-Collins oral history interview, 75.
50. Ibid., 31.
51. "Remembering the Space Race," 11.
52. Ibid., 11–12.
53. Banow-Collins oral history interview, 69.
54. Ibid., 22.
55. Joel Banow, telephone interview with the author.
56. Banow-Collins oral history interview, 13.
57. "Remembering the Space Race," 19.

58. James, *Walter Cronkite*, 130.
59. John J. O'Connor, "TV: Schirra and Cronkite again Team for Apollo 15," *New York Times*, August 4, 1971, 67.
60. Ibid.
61. Banow-Collins oral history interview, 31.
62. Ibid., 56.
63. Ibid., 46.
64. Ibid., 52.
65. Ibid., 41.
66. Ibid., 43.
67. Ibid., 30.
68. Ibid., 58.
69. Paul Haney, *Newsweek*, July 28, 1969, 29.
70. Ibid., 12.
71. Banow-Collins oral history interview, 31, 37.
72. Shoemaker and Reese, *Mediating the Message.*
73. Ibid.
74. Ibid.
75. Stuart Hall quoted in Shoemaker and Reese, *Mediating the Message.*
76. Banow-Collins oral history interview, 9.
77. Ibid., 10.
78. "Remembering the Space Race," 20.
79. Ibid.
80. Ibid., 5.
81. Joel Banow, telephone interview with the author.
82. Ibid.
83. Ibid.
84. Ibid.
85. Ibid.
86. Ibid.
87. Banow-Collins oral history interview, 49.
88. Ibid.

Part II

THE RHETORICAL STRUCTURES, 1968–2022

As discussed in part I, NASA's Public Affairs Office and CBS's news constraints and organizational intricacies influenced the reporting of one of the most-watched events in the history of humankind. The ways in which these factors influenced the rhetorical strategies of the end product—the televised representations—will now be examined. But the story did not end in 1972 with the completion of the Apollo program. It has continued to be part of America's ritual of celebrating its technological successes in space exploration, though the discourse has shifted over the years, as will be evident. Apollo stories have morphed from the time they were initially broadcast to the way they are remembered today, just as media technology has changed. As Michael Schudson has noted, "No part of culture more directly expresses common understanding of the world than does language, and no part of culture is more widely shared. Language is what we think with."[1] Thus, my exegesis of televised representations of Apollo on CBS, CNN, THC, and PBS fell into four rhetorical categories defined in chapter 2: nationalism, romanticism, pragmatism, and technology. In sum, while nationalism encompasses references to national pride and prestige, romanticism emphasizes idealism and intangible rewards, and pragmatism focuses on the practical use of space.[2] The technology category uncovers the role Apollo had in fighting the Cold War while touting America's technological successes and the interjection of entertaining elements to appease audience anxiety regarding high technology. More importantly, it recognizes human technology workers, including those previously underrepresented, and their contributions to Project Apollo. The next four chapters address each of these narrative structures, and the final chapter coalesces these elements to offer a narrative on what the rhetorical shifts mean.

NOTES

1. Michael Schudson, *Watergate in American Memory: How We Remember, Forget, and Reconstruct the Past.* (New York: Basis Books, 1992), 153.
2. Mark E. Byrnes, *Politics and space: image making by NASA* (Westport, CT: Praeger Publishers, 1994), 171.

Chapter 5

Nationalism

Space historians have repeatedly argued that the Apollo program boosted American pride at a time when the public began to doubt big government-sponsored endeavors such as the Vietnam War, civil rights reform, and the War on Poverty.[1] Presidential advisors were also dubious. But despite their warnings, Lyndon B. Johnson decided to move ahead with Apollo for several reasons: he helped create the program; a large number of his congressional supporters lived in states that had been awarded aerospace contracts; and he realized the symbolic meaning that a successful American achievement would have for public opinion.[2] NASA recognized it as well: Administrator James Webb argued for "a plan that would aim at enhancing national prestige."[3] CBS therefore followed suit: often reporting from "America's Moon Port" (CBS, 10/9/68), the network reinforced the notion that the Apollo program was an "American" adventure that belonged to the people. To illustrate, it resurrected several patriotic symbols, including the American flag, images of the late President John F. Kennedy, and framed the United States in terms of its place in history. These icons of pride and success have continued to be repackaged on CBS, CNN, THC, and PBS in post-Apollo years.

OLD GLORY AND OTHER AMERICAN SYMBOLS

As expected, Apollo broadcasts displayed the American flag prominently. While astronauts installed Old Glory on the lunar surface (see figure 5.1), Mission Controllers and members of the public waved it as television cameras recorded the images of a proud and triumphant nation. The intent behind invoking the icon fell into three categories. First and most obvious, the flag conjures up images of American pride at meeting a goal. This was especially

Figure 5.1 American Flag Planted on Moon during Apollo 11 Mission. ***Source:*** **Courtesy of NASA.**

evident during the first part of the Apollo project. Reliance on the icon at the beginning of the program with Apollo 8 continued through Apollo 11, the *ne plus ultra* of the program. Here, symbolic of what David Schoumacher called "the traditional evidence that a challenge had been overcome" (CBS, 7/29/69), it also stood as a symbol of "inspiration," as Chicago Mayor Richard Daley announced at a parade honoring the astronauts. "No one was more inspired than the hundreds of millions of Americans when that flag was planted on the Moon" (CBS, 8/13/69). The City of Chicago painted the normally yellow street lines red, white, and blue for the parade—the same city that had witnessed some of the most violent rampages in its streets just a year prior at the Democratic Convention where protesters voiced opposition to the Vietnam War. Now, riots of flags waved by thousands of cheering Americans drove the point home that America was at last victorious.

In addition, Cronkite reminded viewers that the House had passed a $3.9 billion annual authorization bill for NASA, including a "prohibition against American astronauts planting any but the United States flag on the Moon" (CBS, 6/10/69). It was evident that the Moon landing was solely a "U.S.

achievement" (CBS, 7/15/69), even though "we came in peace for *all* mankind." Just two days prior to liftoff, CBS News Commentator Eric Sevareid, quoting British novelist Virginia Woolf, said, "They, the Americans, have a light that dances in front of them. They face the future and not the past" (CBS, 7/14/69). As CBS asserted, that dancing light was now the Moon, and it belonged to the United States. The Stars and Stripes, clearly the network's symbol of choice, was linked with "teamwork," hinting that together Americans could accomplish anything. As astronaut Eugene Cernan noted, "[It's] really hard to believe, but I am convinced now more than ever that there's no place we can't go and nothing we can't eventually do because of the talent that we've got in this great country of ours" (CBS, 5/27/69). CBS followed up with clips of NASA's Mission Control personnel cheering and waving American flags.

Pride obtained from meeting a goal was repackaged years later to remember Apollo's success, even as reporters spoke of current NASA technological undertakings. For instance, as CBS's John Roberts reported that NASA's Lunar Prospector would reveal if water had existed at the base of the Moon, cameras cut to vintage footage of two astronauts planting the U.S. flag on the lunar surface (CBS, 7/29/99). Such images underscored the notion that exploration of any kind—human or mechanical—is "American." THC, as usual concentrating on the nuts and bolts of Apollo's technologies, also trotted out the flag, which functioned as a means to say that the technology worked *because* it was American. The usual reprise of Mission Control members cheering and waving the red, white, and blue as Armstrong's black-and-white image descending the ladder onto the lunar surface competed with flag patches on uniforms and spacecraft. As PBS put it, because "the American taxpayer had footed the bill, the flag came first" (PBS, 1/25/78). Feelings of pride were also conjured up when CNN's John Zarrella announced the death of Alan Shepard, the first American astronaut in space, who had been an Apollo astronaut. With the flag in prominent view, Zarrella explained how Shepard "carried the burden of restoring national pride" as the first American in space (CNN, 7/22/98).

Despite the sometimes bathetic rhetoric, such as President George H. W. Bush's inviting viewers to "step outside tonight with your children, your grandchildren, lift your eyes skyward and tell them of the flag—the American flag—that still flies proudly on the ancient lunar soil" (CBS, 7/20/89), it should be mentioned that, in post-Apollo years, the networks seemed to be more creative. Flags proliferated in THC's unique camera shots, as audio described the Moon as having a "stark beauty of its own, much like the high desert of the United States" (Armstrong) and "magnificent desolation" (Aldrin) in *Modern Marvels: Apollo 11* (THC, 7/21/04). THC's flag registered red, white, and blue against the black backdrop of space, as if to claim all of space for the United States. Old Glory appeared again as the narrator described the lunar samples collected for scientific experiments. The Moon seemed to have belonged to America. Additionally, PBS's *One Small Step* focused on the letters "USA" juxtaposed with the flag on the Saturn V rocket as it moved upward. The rocket's massive, stately motion created the illusion of rising Stars and Stripes. Clearly, the networks were exploring new ways to incorporate the flag apart from the standard image of two astronauts planting the emblem on the Moon.

Secondly, the flag conveyed international connotations, especially in later years as the United States has had to rely on foreign assistance to continue space exploration. Astronaut Mike Collins once said that when future astronauts "step out onto the surface of Mars or some other planet, just as I listened to man step out onto the surface of the Moon, I hope I hear him say; 'I've come from the United States of America'" (CBS, 9/16/69). Such sentiments would abate years later when the need to cooperate with other nations in joint ventures would become necessary to offset the monumental costs of space exploration. Old Glory, therefore, continues to be featured as a way to tout the United States even though today's space exploration requires funding from international and private enterprise sources. In news reports post-Apollo, the flag continued to be featured prominently with each remembrance of the Moon missions.

Finally, placing the U.S. emblem on the lunar surface symbolized a massive technological feat mounted by ingenious American engineers (figure 5.1). Not only did Apollo signify the actualization of getting to the Moon, but also the engineering and logistical challenge of transporting and erecting an object—the flag—in zero gravity. Uncertain of the materials that could survive the lunar atmosphere, NASA engineers circumvented a number of engineering constraints: what could withstand the climate of an infinite number of years? How could the flag appear to wave with no wind? As payload to the Moon, what could its maximum weight be, and how would the astronauts assemble it once there, given their pressurized space suits and gloved hands?[4] Similarly, the boot print from Aldrin's space suit (figure 5.2) served as a symbol of human ingenuity, and the networks post-Apollo used it in conjunction with stories of Apollo's victorious technology. While the footprint signaled that man was there, the flag reminded viewers that it was an *American* man.

THC welded the flag and the rocket to tow other distinctly "American" icons: the Statue of Liberty and baseball. THC's *Crisis in Space: The Real Story of Apollo 13* and *Modern Marvels: Apollo 13* rhapsodized the Saturn V rocket as "taller and heavier than the Statue of Liberty" (THC, 1/22/01). Here, the symbol of freedom spoke of the courage, technology, and power of the United States. PBS's *To the Moon* juxtaposed a computerized animation of the "monster" Saturn V rocket with a computerized animation of the

Figure 5.2 Apollo 11 Astronaut Footprint. *Source:* Courtesy of NASA.

Statue of Liberty to establish the difference in scale (figure 5.3). Shortly following, Lovell retold the story of the Moon being made of *American* cheese (THC, 7/13/99), but cheese obviously could not resonate, so THC's *Triumph and Tragedy: Apollo 11 and Challenger* narrator Mike Wallace retreated to baseball metaphors to describe how the rocket was precisely targeted to a predetermined spot. Launching Apollo was like "firing a baseball from Dodger's Stadium in Los Angeles to home plate in the Yankee Stadium in New York" (THC, 3/10/99). In addition to the significant geographic distance between the teams, it is interesting that Wallace chose the most storied rivals in all of baseball, having met more times than any Major League Baseball teams at 11 times in the World Series,[5] perhaps playing up the U.S. vs. USSR rivalry. Space exploration, then, was packaged as American as baseball.

Kennedy and the Space Race

Discussing the space race and John F. Kennedy in tandem congealed a nationalistic meaning. The race metaphor naturally placed the United States ahead of the Soviet Union while reminding viewers that it was the martyred president

Figure 5.3 Saturn V Rocket "Taller than the Statue of Liberty." ***Source:*** **PBS NOVA To The Moon.**

who issued the challenge that would give America the edge. CBS played up the contest. From the onset, it reassured viewers that the United States stood a good chance of being first on the Moon. Reporter Steve Rowan claimed,

> [NASA Administrator] Dr. Thomas Paine doesn't believe the Russian achievement makes them automatically winners of the race for the Moon. In fact, while admitting that [the Russians] have done something the U.S. cannot do at the moment, he suggested that the U.S. astronauts still might be first on the lunar surface. (CBS, 9/23/68)

In other words, the United States was still number one even if it was lagging behind the Soviets, for "man, American man, would be on the Moon before this decade [was] out," according to Cronkite (CBS, 10/22/68). Looking ahead to Apollo 8, the first mission to circle the Moon and return to Earth, Cronkite called the race "an urgency" (CBS, 10/9/68) because the Soviets might beat us in a *manned* trip around the Moon.

This "race" metaphor also created a drama that would keep Americans "tuned in" until the climax, which of course came with Apollo 11. Here, CBS reminded Americans of what first took them to the Moon: competing ideologies. Over a picture of the Saturn V rocket posed for launch on July 16, 1969, CBS's audio echoed Kennedy's challenge:

> We choose to go to the Moon in this decade and do the other things. Not because they are easy, but because they will serve to organize and measure the best of our

> energies and skills, because that challenge is one that we're willing to accept, one we are unwilling to postpone, and one we intend to win. (CBS, 7/16/69)

Kennedy had intended Apollo to be a race. To intensify the contest, this clip was juxtaposed with a segment featuring "man in street" interviews, a news ritual that unites Americans into one ideological community. Reporters gave a few of the estimated 528 million people in 33 countries who watched the televised launch a chance to voice their opinions of Apollo 11 (CBS, 7/16/69). Said a male viewer in London: "I'd like to see the West Side beat the Iron Curtain Boys" (CBS, 7/16/69). The reference was to the popular "West Side Story" Broadway play, a story of rival gangs, the Jets and the Sharks. Pitting nation against nation created the sense of community of which Carey has spoken: by "taking sides," the viewer therefore feels part of the process and is enabled to live vicariously through NASA's handpicked talent. Hence, the viewer comes to support the goal set forth by NASA and CBS.

Framing Apollo as a race has endured. Mission Control Director Gene Kranz explained why:

> The Kennedy speech really hit home to a lot of us. I believe this was a solid opportunity to demonstrate once and for all who was the leader of the free world and I wanted our nation to be the leader. (CBS, 7/14/94)

On the 30th anniversary of Apollo 11, clips from Kennedy's speech preceded CBS's Richard Schlesinger's remarks that "the goal wasn't just to get to the Moon; it was to get there before the Soviets. The high frontier was the new frontier in the Cold War" (CBS, 7/20/99). Apollo, in short, burnished the legend of America's global leadership. As Ernest Gellner has noted, "Nationalism usually conquers in the name of putative folk culture,"[6] and CBS seemed to have appointed Kennedy as the mouthpiece of U.S. lore.

The networks have also used the Apollo story as an occasion to illustrate why U.S. ideology reigned supreme. During the Apollo years, for example, Sevareid called the Apollo 11 mission a "feat" (CBS, 7/21/69) compared to the Russian Luna 15 that flew to the Moon simultaneously with Apollo 11. As Luna 15 flew, Cronkite reported that scientists listened to Apollo 11 signals in one ear and Luna 15 signals in the other (CBS, 7/17/69). A week later, reports indicated that the unmanned Russian satellite crashed into the Moon. CBS News seemed to take pride not in the fact that the Luna flight had failed but rather that the Russian people learned of it through "inconspicuous stories buried in Moscow's newspapers" (CBS, 7/22/69). The implication was that "our ideology is better than theirs" because an open society is superior to a closed one. In any case, covering the two stories was testament that media still saw the mission as a race with the Soviets.

Stories of a transparent society have also endured, which is not surprising given that most made-for-television movies and documentaries focus on the "idea of the American experience rather than the process of historical change over time."[7] THC's *Failure is Not an Option* and PBS's *One Small Step* acknowledged America's openness even though its technology had failed several times before outpacing Russia's. America's rocket failures, "symbolic of national weakness" (THC, 8/24/03), had momentarily given the Russians a technological leadership that by rights belonged to the United States. To dramatize, THC listed a host of Russian technological firsts, then cut to Kennedy at the podium warning Americans that the United States was behind. The narrator recalled the nation clutching at straws: commentators could claim that, although Alan Shepard's suborbital flight did not match that of Soviet cosmonaut Yuri Gagarin in duration, at least it was televised live for the world to see. Here, vintage footage of television cameras captured the moment at the launch site while Mission Control fretted over disaster and as the world watched. Television—called the "expanding vista" in the Kennedy years[8]—thus inserted itself into the Apollo saga as an agent of a free and open society: the United States might have been behind technologically, but at least it was a democracy, transparent to the all-seeing lens. "The Russians [were] still ahead, but they fl[ew] in secret. Launching Shepard on live television [gave] America an edge in the battle for world opinion," the THC narrator observed as the footage traversed to President Kennedy pinning a Medal of Honor on Shepard's coat after the flight (THC, 8/24/03).

The handsome icon also served as a paradoxical symbol of all that was good and all that was tragic in American life. While Kennedy's image was invoked to explain Cold War tensions, he, although assassinated, represented America overcoming a challenge. Whether by design or accident, when astronauts later landed on the Moon, CBS News featured audio of them exclaiming, "Absolutely incredible! Wow! Camelot! Right on target!" (CBS, 12/11/72). Camelot and Apollo seemed naturally paired, for the Apollo tale, like Camelot, was replete with "heroes, fairy tales, [and] legends," what White said "history was all about."[9] Because the Kennedy administration became known as "Camelot"—a "magic moment in American history"[10]—the Camelot reference may have implied that the Moon was the new utopia awaiting Americanization, an idea begun by America's tragic but shining knight. Either way, U.S. ideology, through images of Kennedy, toppled U.S.S.R. ideology.

Kennedy's ringing voice was also asserted in stories that the United States could overcome its challenges, and his message provided the test for subsequent ventures. Like the flag, Kennedy had to be trotted out to confer legitimacy on space events. For instance, when President George W. Bush announced another lunar mission, CBS's Dan Rather promptly ran a clip of Kennedy saying, "we choose to go to the Moon . . . and do the other . . . Not because they are easy . . ." (CBS, 12/5/03). As John Roberts explained, "President Bush is preparing a dramatic announcement . . . a JFK-style call for reinvigorating American's space program by saddling the high frontier . . . to make a big election year splash" (CBS, 1/9/04). Calling the plan a "Kennedy moment for Bush" in an election year, Bill Plante further explained that another lunar excursion and a mission to Mars were perhaps ways to revitalize NASA—to give it goals—despite public skepticism that manned missions were necessary at all (CBS, 1/9/04). Just as Apollo had aimed to beat the Russians, said Plante, Bush's proposed missions would also be driven by national security concerns. China had just recently launched a man into orbit and had also hinted at a permanent Moon base (CBS, 1/14/04). Oddly, CBS did not take note of Apollo 11's 35th anniversary, but CNN did, and predictably rehearsed the politics of the Cold War. History seemed to be repeating itself. Just as Apollo mitigated failed efforts in Vietnam, clearly Bush was hoping for post-9/11 political capital to offset public resentment of involvement in Iraq.

For Apollo 11's 50-year anniversary coverage, CBS depended on Kennedy's legacy when its stories needed to convey that goals had been met, but also to remind viewers of how important it is to continue space exploration. For instance, although the launch date was July 20 and the Moon landing was July 21, CBS anchor Norah O'Donnell recognized July 24 as the date that "Kennedy's challenge of returning men safely to earth had been fully realized" and as "a mission accomplished" (CBS, 7/24/19). Accompanying her words was footage of Kennedy delivering his famous "Before this Decade is Out" speech, juxtaposed with the American flag being positioned on the Moon, followed by splashdown coverage. Conversely, a few days before that broadcast, O'Donnell sat down with billionaire entrepreneur Jeff Bezos and Kennedy's daughter, Caroline Kennedy, who stated that her father saw space exploration as something that would "advance freedom and benefit humanity," as opposed to winning the Space Race, while the standard footage of JFK and the flag rolled in the background (CBS, 7/16/19).

HISTORY AND TIME

The networks employed an additional way to make Americans feel close to and part of the Moon shots by discussing the Apollo story in terms of "being first" and being historical. CBS continually emphasized the United States' firsts, knowing that Americans have long taken pride in making history. Apollo 8 circled the Moon first; the first discovery of volcanoes on the Moon came with Apollo 10; man was first on the Moon with Apollo 11. The Apollo

12 mission televised the first color pictures broadcast live from the Moon, while Apollo 15 featured the first Moon buggy.

Even Apollo 17, the last of the missions,[11] was the first night launch and "the beginning of space exploration" [moving beyond the Moon to other space endeavors] (CBS, 12/4/72, 12/7/72). This glorification of American firsts continued after the apex of Apollo 11, though it seemed CBS had handed over its relentless "cheerleading" rhetoric to Nixon. As with all missions, CBS's ritual of featuring the U.S. president's reaction to the mission would be included in the evening news. Nixon predictably responded:

> America, the United States, is first in space. We're proud to be first in space.[12] We don't say that in any jingoistic way. We say it because as Americans we want to give the people of this country, and particularly our young people, the feeling that they're here and an area we can concentrate on as a positive goal, concentrate and be proud of being Americans, be proud of what we've accomplished, not only for ourselves, but for future generations, and for the whole world. (CBS, 11/14/69)

According to Dorothy Nelkin, science and technology in the United States have traditionally been explained by the media with a focus on "firsts" for those very reasons: pride and leadership.[13]

CBS also consistently described the astronauts as "writers of history." When the Apollo 8 crew emerged from the dark side of the Moon, Cronkite called their reappearance an "historic confirmation" of American know-how (CBS, 12/25/68). Similarly, the day after the successful Apollo 10 flight ended, Cronkite referred to the astronauts as "three who wrote yet another bright page in conquering the cosmos" (CBS, 5/27/69). But the astronauts, in a post-flight news conference, followed their own script: "Through this endeavor, we hope that we've made you and millions of people of the world more a part of the history that's being made in our day and age" (CBS, 5/26/69). Here, the astronauts enabled the American people to become the creators of history. Again, participants in culture are more likely to believe in and preserve its rituals. Newsmen seemed to have mimicked the astronauts: in the "age of technology" (CBS, 7/11/69), perhaps the most telling description of history, however, was in Cronkite's report of *how* history was being made. With the help of a cameraman, he focused just as intensely on the hundreds of thousands of ordinary citizens gathered at Kennedy Space Center to witness "history in making." In his folksy manner, Cronkite reminded audiences that it was the people for whom history was being made, and it was history that was being made in the most ordinary and American of ways.

> History was here. History observed in shirtsleeves and bikinis, history seen from little rented sailboats in the Banana River, and from the tailgates of camper

wagons in Titusville. History viewed through a haze of smoking charcoal grills. For a half a million Americans, staying home and watching was not enough. (CBS, 7/16/69)

That Americans wanted to witness the Apollo 11 mission as it launched spoke to the relevance it had in their lives, like the parents who woke their children in the middle of the night to see the Eagle touch down on the Moon, or those who could not travel to Florida but instead took photographs of their television sets because they knew it was a significant moment. Thus, not only did CBS appeal to Americans' sense of history, but it also made Americans a product of that history by involving them. CBS eloquently laced context into the Apollo 11 mission: just before launch, Sevareid described Apollo as "just the beginning" and "a new stage in the evolution of [man's] species," comparing the landing on the Moon to "the first crawling of the first amphibian creature out of the primeval swamps under dry land" (CBS, 7/14/69). For Cronkite, Apollo 11 was a turning point: "50,000 years of human history [was] about to turn the corner" (CBS, 7/15/69), as it would be "one of the towering events in our history" (CBS, 7/17/69). To create a sense of community by linking Americans together as witnesses, CBS devised its own vernacular to chronologically mark events: "the lunar age" (CBS, 12/24/68) and "the beginning of a new era where man understands himself" (CBS, 8/14/69). Sevareid even suggested calling the pre-launch era "B.M." ("Before Moon") (CBS, 7/17/69), in keeping with the tradition of marking a date "A.D."

But as a technological endeavor, CBS also placed the mission in historic context by comparing it with other important technologies that would surely resonate with viewers. Calling the Saturn V the "300-ton stack of rocket" (CBS, 7/8/69), Cronkite said it was a "vessel which will rank in history with, perhaps before, Lindbergh's *Spirit of St. Louis*, Darwin's *Beagle*, and Columbus' *Santa Maria*" (CBS, 7/10/69). The network also used metaphors to describe the command module named *Columbia* (after Jules Verne's *Columbiad* in his 1866 novel of space flight, *From Earth to the Moon*), referring to it as "the historic vessel" (CBS, 7/17/69). The name *Columbia* was selected to honor Christopher Columbus. One need not know much about history to understand the significance of such names. Even the ship that retrieved the astronauts at splashdown seemed historically significant: the *Yorktown*, called the "Fighting Lady of World War II," was named for the Battle of Yorktown that marked America's defeat of the British in the Revolutionary War in 1781.

In keeping with that tradition, networks in later years put fresh spins on the old story—again, to mark the event historically, but also to spice up the narrative. *Triumph and Tragedy: Apollo 11 and Challenger*, for example, also added a political dimension. Mike Wallace described how the aircraft carrier *John F. Kennedy* was scheduled to recover the astronauts after splashdown in the ocean when they returned to Earth. The Nixon administration, however, "had no intention of reminding the American people of Apollo's beginning" (THC, 3/10/99), so instead *USS Hornet* rescued the astronauts from the waters. According to Wallace, "Despite the space programs' close identification with presidential rival Kennedy, President Nixon did not pass up the chance to welcome the astronauts home" (THC, 3/10/99). The network then cut over to footage of Nixon telling the astronauts that, because of Apollo 11, it had been "the greatest week in the history of the world since the creation" (THC, 3/10/99). THC could not resist providing subtext: a lesson in presidential history of the Kennedy/Nixon rivalry and an example of a paranoid Nixon.

History was revisited on network news channels with each anniversary of important Apollo mission dates, as well as with the passing of each Apollo astronaut. For example, in *CBS Evening News* coverage of Apollo 8's 40-year anniversary, Charles Osgood reminds us that "Their voyage offered hope to a weary world," while presenting a new spin on old historical tensions: "NASA improvised the trip at practically the last minute after getting erroneous word that a manned Soviet Moon mission might be imminent. Christmas Eve, 1968," as CBS provided old footage of the Genesis reading and Earthrise photos (CBS, 12/21/08). Five years later, CBS's Dean Reynolds stated, "Three humans circled the Moon for the very first time and briefly connected the people of this planet" as the Earthrise photo was shown, a photo that "went viral before there was a going viral" (CBS, 12/24/13). Similarly, for Apollo's 50th anniversary, CBS reminded viewers that the Apollo 8 journey "marked a triumphant way to mark a traumatic year," making it all the more "momentous" as it showed old news footage of the Vietnam War, political assassinations, and uprisings. Apollo 13 astronaut Jim Lovell said, "We finally brought back something Americans could be proud of," as the Christmas broadcast and Earthrise were again brought to memory by correspondent Chip Reid (CBS, 12/21/18).

In a similar fashion, when Apollo 11's Neil Armstrong died, CBS paid tribute to this "reluctant American hero who shunned the spotlight even though he played the leading role in one of the greatest moments in American history" (CBS, 8/25/12). Two nights later, CBS found a new angle on Armstrong's story, which was that the Moon landing almost did not happen as it recalled the problems with the lunar lander's alarm system. Mission Control Director, Gene Kranz, gave credit to Armstrong for landing on the Moon. "This was exciting. It was risky, but it was typically America. You know, this is what Americans can dare Americans can do," as the network showed footage of Armstrong and Aldrin planting the Stars and Stripes on the lunar surface. (CBS, 8/27/12).

Finally, so-marking Apollo fueled much-needed encouragement in later years to justify returning humans to the Moon. While *To the Moon* depicted the rocket's launch pads as "monuments" that "loom[ed] like some 20th Century Stonehenge" (PBS, 7/13/99), *Triumph and Tragedy: Apollo 11 and Challenger*'s Mike Wallace put Apollo into historical perspective for viewers.

> No Earthly troubles could diminish the majesty of what had been achieved. It seemed that this was the beginning of a great new age of space exploration. And if history had gone a different way, if there had been more money to go around, if there had never been a Vietnam, no struggles in American streets, no Watergate scandal, no growing doubt about the technology itself, manned space exploration of space might have remained a high priority. (THC, 3/10/99)

Evoking this kind of nostalgia was intended to prod viewers to agree that space exploration should remain a national priority. As Mission Control's Steve Bales said, it would be difficult for a historian to explain why the United States did not go back to the Moon for 30, 40, or even 50 years (PBS, 7/13/99). The space exploration continuation narrative appeared in various stories on *CBS Evening News* into the next century as well. On Apollo 11's 40-year anniversary, the narrator said that the anniversary celebration had been tempered by disappointment that America in four decades since had not probed deeper into the unknown. Buzz Aldrin believed the United States should be the first to land on Mars. Meanwhile, Apollos 10 and 17 astronaut Gene Cernan told the network, "The Spirit of Apollo has been lost. I really believed that we'd be back to the Moon by the end of that decade and on her way to Mars by the turn of the century" (CBS, 7/20/09). The narrator continued, "For America, the Moon landing was the payoff. Aldrin says it should have been the beginning" (CBS, 7/20/09). One reporter, Jeff Glor, even joked that "U.S. astronauts are going to be forced to hitch rides on Russian Soyuz rockets" following the report that then-President Barack Obama had canceled the $110 billion Constellation program, for which "lots of space veterans were not happy about" (CBS, 4/14/10). If historic perspectives failed to convince the public to support future missions, perhaps appealing to its romantic sensibilities could.

NOTES

1. See, for example, Launius, McCurdy, and McDougall, to name a few.
2. Howard E. McCurdy, *Space and the American Imagination* (Washington, DC: Smithsonian Institution Press, 1997), 84.
3. Ibid., 97.
4. Anne Platoff, "To Go Where No Flag Has Gone Before," *NASA Contractor Report 188251*. NASA History Office, Washington, DC, 1993.

5. The two teams had played one another in the World Series 11 times when Wallace made the comparison in 1999. Since then, the Yankees and Dodgers have met a twelfth time, in the 2024 World Series.

6. Ernest Gellner quoted in John Bodnar, *Remaking America: Public Memory, Commemoration, and Patriotism in the Twentieth Century* (Princeton, NJ: Princeton University Press, 1992), 11.

7. William Cohn, "History for the Masses: Television Portrays the Past," *Journal of Popular Culture* X, no. 2 (Fall 1976): 281.

8. Mary Ann Watson, *The Expanding Vista: American Television in the Kennedy Years* (New York: Oxford University Press, 1990).

9. Theodore White, "Camelot, Sad Camelot," *Time*, July 3, 1978, 47.

10. Ibid.

11. See, for example, *CBS Evening News* broadcasts on December 4, 5, 6, 14, and 19, 1972.

12. It should be noted that Nixon misspoke. The United States was not the first in space; the Soviets were. Nixon was referring to being the first on the Moon, or possibly the first to return humans to space repeatedly and successfully.

13. Dorothy Nelkin, *Selling Science: How the Press Covers Science and Technology* (New York: W. H. Freeman and Company, 1987), 51.

Chapter 6

Romanticism

In her social history of American technology, Ruth Schwartz Cowan traced the roots of romanticism to 1780s Germany. It spread to Britain in the early 1800s and then to the United States by the mid-nineteenth century. Romantics "celebrated passion at the expense of reason" and prized "spontaneity [over] practicality."[1] Initially, poets, artists, and musicians were opposed to industrialization and all it represented: science, technology, and progress. Cowan marks the Civil War as a turning point for devout romantics, about the time that inventors such as Whitney, Morse, Edison, and Bell became national celebrities, praised for inventions that liberated Americans. Thus, one could hail progress and still be a romantic, for "industrial technology would create prosperity and material comfort for ordinary people, a set of ideas that is sometimes referred to as the American Dream."[2] A century later, NASA and television networks needed a way to blend romanticism with technology, which was imperative in their efforts to tout the Apollo program. How, for example, would they engage viewers in an expensive government-sponsored endeavor replete with gargantuan machinery that would uncover the mysteries of craters, boulders, and meteors that pockmarked the gray, dull surface of the object of their goal?

Romanticism, as defined by Mark Byrnes, is an image that emphasizes the spectacular while enabling the public "to escape temporarily the humdrum routine of daily life."[3] Pertaining to NASA's endeavors, romanticism addresses emotional rewards such as overcoming challenges, inspiring people to achieve greater goals, obtaining opportunities to explore new frontiers and satisfy curiosity, and having the imagination to dream and exercise spirituality in a peaceful world. Taken further, romanticism, as it was portrayed on television, involves the efforts of heroic people (astronauts and members of Mission Control) and provides the kind of excitement that enables the

American public to live vicariously through these characters who are so skillfully crafted by NASA. While the agency has traditionally depended on romantic images throughout its existence, they were most heavily utilized during the Apollo years. And television, like no other medium, can deliver romantic narratives through its ability to visualize and verbalize mythic characters, to play and replay events to the point where they become part of the cultural fabric. As Walter Cronkite—the man Katie Couric said "took us there and back" (CBS, 7/20/09)—indicated during Apollo's 40th year coverage, "At a time when there were a lot more problems in this country and elsewhere where people were downcast, everybody there [at NASA] was upcast and we were looking toward the stars, looking toward the Moon" (CBS, 7/20/09). The meanings of Apollo, then, are produced through the stories that television continues to tell.

EXPLORATION AND FRONTIERISM

One such story that has appealed to American culture is that of a pioneering spirit. "The urge to explore is innate in man," according to NASA, and "space has been, and will continue to be an American frontier."[4] Conquering new frontiers begets innovation and fosters discoveries,[5] and CBS capitalized on the recurring romantic narrative that evoked exploration and the Western frontier. As historian Roger Launius has noted, repeated use of the frontier metaphor has convinced the American public to support spacefaring activities because it embodies American ideals such as democracy and optimism.[6] Following NASA's rhetoric, the networks employed several strategies to incorporate this narrative. One was to cast the astronauts, and then engineers, as modern-day Christopher Columbuses and frontiersmen. For instance, Eric Sevareid contrasted the three Apollo 8 astronauts with Columbus. Unlike those of Columbus's time, modern-day explorers, thanks to television, could "perhaps change us as perceptions of Europeans changed with new knowledge of the New World to the West" (CBS, 12/23/68). In addition to being called "foreign visitors to a new land" (CBS, 5/19/69) and "first voyagers to another world" (7/9/69), the astronauts' "incredible voyage" (CBS, 7/16/69) enabled them to be "conquering heroes" and "pioneers of the past" (CBS, 8/13/69). But there were also several references to future frontiers. While Spiro Agnew predicted that Apollo 11's success would enable man to explore Mars by the end of the century (CBS, 7/16/69), an unknown person on the street interviewed by CBS additionally forecasted U.S. excursions to Venus and "beyond" (CBS, 7/25/69). If man could successfully land on the Moon and return safely home, then he could venture on to other planets as a natural progression of exploration.

Years later, networks credited the providers of technology with being America's explorers and frontiersmen as much as they initially did the astronauts. *Failure is Not an Option*, for example, focused on Apollo engineers, offering the riveting origins of Mission Control, a "revolutionary concept that enabled flight control from the ground" developed by Christopher Columbus Kraft during a time that preceded global satellite communications (THC, 8/24/03). While broadcasters such as PBS merely referred to him as Christopher Kraft, THC's use of Kraft's middle name served an obvious purpose. Like the astronauts, these engineers were pioneers in their own right. Most were the first in their families to realize the American dream of going to college. As with astronaut jobs, NASA did not need to advertise to fill engineering positions; it received thousands of applications annually. According to newscasts, like the astronaut pool, Mission Control engineers were a homogeneous collection of white men who graduated from college within approximately one year of one another.

Just as CBS focused on the astronauts' backgrounds during the Apollo missions, *Failure is Not an Option* featured the humble beginnings of the engineers of Mission Control, focusing on one engineer in particular, John Aaron, whose fellow engineers called "the most capable flight controller ever to work in NASA's Mission Control" (THC, 8/24/03). Clearly playing up the rustic pioneer type, THC featured Aaron in a cowboy hat amid rugged mountains and a seemingly endless blue Texan sky as the backdrop while Aaron told the interviewer he was trying to build up a herd of heifers. He thought he wanted to be a cowboy (figure 6.1). At this point, the camera panned down his legs to his feet, presenting the viewer with a close-up of Aaron's cowboy boots for several seconds. Aaron laughingly explained what happened next in his soft Southern accent:

Figure 6.1 John Aaron, NASA Mission Control's cowboy. *Source:* The History Channel, "Failure is Not an Option."

> A friend says, "why don't you send in an application?" So, the next thing I knew they didn't even grant me an interview. They sent me a job offer for more money than a country boy had ever seen, by the way. (THC, 8/24/03)

Thus, the network presented viewers with another cowboy—one who might *not* have flown into space but instead was skillfully responsible for getting the astronauts to and from the Moon safely.

The "country boy" icon fits the space program nicely. It suggests "the 'bedrock' American values of solidity, respect for authority, old-time religion, homebased virtues, and patriotism,"[7] as well as the "loyalty, honor, and faithfulness of working people while also emphasizing the self-reliance of the individual and the family as the ultimate source of security."[8] Values of "country boys" mimic what has been illustrated here: the networks and NASA portrayed the astronauts and later the providers of technology to be faithful, hard-working, God-fearing, individualistic, and rugged country boys to sell the program to Middle America. This was also evident in less subtle ways: documentaries such as PBS's *To the Moon* featured country music in the background as it pictured the astronauts training in merciless territories (out West, of course), soon to face the challenges of the unforgiving terrain of the Moon (PBS, 7/13/99). Astronauts seemed to be fans themselves. Pete Conrad was inducted into the Country Music Hall of Fame for promoting country music during the Apollo 12 mission and other astronauts would follow suit in later missions.[9] The networks, then, illustrated that the astronauts, though "the kind of guy[s] you want to sit down and have a beer with" (THC, 7/21/04),[10] were not the only space cowboys during the Apollo years.

The cowboy narrative followed CBS and Apollo narratives into the next century. On July 20, 2021, the 52nd anniversary of Apollo 11, Norah O'Donnell described a "new era of space tourism" in covering billionaire Jeff Bezos's Blue Origin successful launch into space on the first unpiloted, fully civilian suborbital flight, with the youngest and oldest humans ever to fly into space at the time. "For Jeff Bezos, the microgravity dream he first imagined as a five-year-old watching the Apollo 11 Moon landing, was 52 years ago today." CBS rolled the "one small step for [a] man" footage from Apollo 11 before cutting to Blue Origin footage and Bezos wearing a cowboy hat (figure 6.2). "It felt so normal and natural, almost like we humans were evolved to be in zero gravity, which of course is impossible. But it felt that way" (CBS, 7/20/21). Each of the crew members, including female aviation pioneer Wally Funk, received astronaut wings on this historic day. Interestingly, CNN only mentioned Blue Origin and SpaceX in relation to Apollo when it briefly addressed partisan funding solutions for NASA, as it would likely be the most economical way to get to Mars and settle on the Moon, according to astronaut Michael Massimino (CNN, 7/20/19).

Figure 6.2 Jeff Bezos, Blue Origin's Cowboy. *Source:* CBS News, July 20, 2021.

A second way in which the networks touted the exploration theme was to offer it as "The American Way." Here, the move westward was the metaphor of choice. *One Small Step,* for instance, indicated that the race to the Moon was borne out of a "competitive drive that has moved Western people around the Earth for a thousand years or so" (PBS, 1/25/78). Surprisingly, exploration and frontier metaphors were not employed by CBS until the 1990s, the decade following Ronald Reagan's ("the cowboy President") tenure. CBS often recognized America's space exploration as "our tradition" and "mankind's duty," especially in rebuttals to critics who said the technology derived simply did not justify the costs (CBS, 12/14/97). In addition to former astronauts such as Cernan describing the United States as a "spacefaring nation," Jim Axelrod introduced viewers to the film *Magnificent Desolation*, a film in which actor Tom Hanks sought "to reacquaint us all with the wonder" (CBS, 9/23/05). According to Hanks, "Once humankind has been someplace and found it entrancing, they always go back." (CBS, 9/23/05). Similarly, on CNN a few years prior, Cernan indicated that it would take a considerable amount of time for people to fully realize the meaning of Apollo, just as it took "a hundred or more years, not 25, before people ever fully appreciated the significance of Columbus's sailing across the Atlantic" (CNN, 12/11/97).

THC's exploration and frontier narratives were equally trenchant. In *Triumph and Tragedy: Apollo 11 and Challenger*, the astronauts defended the United States' decision to go to the Moon using the same defense that was used 40 years prior: exploration, American style. Stated Apollo 11 astronaut Michael Collins, "We're a nation of explorers. We would have never ventured past Jamestown or Plymouth had we decided to render those little colonies perfect before we continued our exploration, and I think the same thing is true today" (THC, 3/10/99). Neil Armstrong, a man for whom public appearances were encumbrances, added, "I am personally delighted, as an individual and delighted for my colleagues and our country that, even if it was a bad idea, we did it," as he and an audience chuckled (THC, 3/10/99). Therefore, even if no useful data were produced, the United States explored for exploration's sake. America went to the Moon because it could.

In remembering Apollo, anniversaries were opportunities to convince the audience that mankind should *continue* to explore. Calling Armstrong's "one small step" adage "a test pilot's courage with a poet's words," CBS reporter Byron Pitts introduced Armstrong, who said, "The important achievement of Apollo was the demonstration that humanity is not forever chained to this planet. Our visions go rather further than that, and our opportunities are unlimited" (CBS, 7/16/99). Again bringing to mind Columbus sailing the ocean blue, *To the Moon* included vintage footage of rocket scientist Wernher von Braun who warned,

> We consider the control of space around the Earth much like . . . the great Maritime powers considered the control of the seas in the 16th through the 18th Century; . . . if we want control of this planet, we have to control the space around it. (PBS, 7/13/99)

As Michael Schudson has argued,

> By repeating a phrase the audience knows to be tied to a particular past moment, quotation may take on the tone of incantation and evince reverence or irreverence, evoking the audience's attitudes toward the earlier event or even attitudes toward tradition itself.[11]

Rehearsing such rhetoric (especially Armstrong's "one small step" mantra) undoubtedly persuaded viewers to believe that, unless we continued space exploration, the United States would lose its status as the world's superpower. This was illustrated in O'Donnell's 50th anniversary of Apollo 11 in her interview with Jeff Bezos, who stated that "space exploration is critical to our survival" (CBS, 7/16/19) before suggesting that all pollution-causing industries be relocated to the Moon, even though the technology to do so is hundreds of years away. He concluded that, "The only reason we can do the things we can do today is because we are standing on the shoulders of giants" (CBS, 7/16/19). Thus, the networks noticeably took advantage of Apollo's anniversaries to garner support for future space endeavors.

Finally, the networks enhanced the frontier narrative by downplaying the costs of exploration, usually in the form of presidential rhetoric chock-full of exploration themes. Bill Clinton, for example, indicated on CNN that space exploration was more "about the ability of Americans and the ability of human beings everywhere to conquer the seemingly impossible . . . conquering new and uncharted worlds . . . with a pioneering spirit" (CNN, 7/20/94).

However, most of the rhetoric pertaining to the cost of such conquering came with the news of George W. Bush's 2004 plan to return to the Moon and on to Mars. According to CNN's Aaron Brown, such a plan was:

> President Bush's answer to those who say the country has suffered an ambition deficit ever since Americans last set foot on the Moon. But it comes at a time when the country also has the kind of budget deficit that would stifle even the grandest ambition. (CNN, 1/14/04)

Reporter John King provided a time frame and a cost breakdown associated with the President's proposed missions while alternating with video clips of Bush's speech: "Mankind is drawn to the heavens for the same reason we were once drawn into unknown lands and across the open sea" (CNN, 1/14/04). Here, the reality of a costly space program was explained by CNN while the romantic, frontier-like speech to sell the program was delivered by Bush. Evoking nostalgia, CNN immediately cut to footage of a younger Cronkite removing his eyeglasses while excitedly exclaiming, "Boy, oh boy!" as the astronauts landed on the Moon. CNN then interviewed Cronkite, who opined,

> July 20, 1969 is, I'm sure, going to be perhaps the one historical date that children 500 years from now will recognize. And why can I be sure of that? I'm sure of that because think back 500 years now. It was a very important age in Europe; things were developing there at some pace. And what is the date that is remembered? October 12, 1492. (CNN, 7/20/04)

Brown ended the newscast by saying, "there will never be another feeling quite like it, a feeling of such discovery and pride" (CNN, 7/20/04). Pairing space with previous expeditions enabled the networks to show by example that exploration is beneficial to Americans, regardless of cost. This was not new to the networks, which had known since their inception that casting American ideals to viewers not only increases audiences but also promotes national goals such as exploring new frontiers. Building on the country's bedrock values doubtless gave Americans something of which to be proud.[12] In sum, while historians have argued that it was through the frontier experience that "America became a land of peace, prosperity, reason and liberty,"[13] media scholars have noted, "The news particularly seems to celebrate individuals who 'conquer nature'."[14]

By 2009, the 40th anniversary of Apollo, CBS reported that the space shuttle Endeavor docked with the International Space Station. *CBS Evening News* anchor Katie Couric announced, "As NASA looks back with pride, Space Correspondent Mark Strassmann tells us it's also looking ahead." Strassmann then goes to the empty Mission Control room in Houston and discusses NASA's Constellation spacecraft, which would enable astronauts to reside on the Moon. He then cuts to Apollo 11's Buzz Aldrin, who says NASA needs to go places people have never gone before, that America has to make hard choices, such as where to go, risks to take, and how much they are willing to spend, as a return to the Moon would be an estimated one hundred billion dollars. Strassmann concluded with "NASA must make the case for space and the next giant leap for mankind" (CBS, 7/17/09). This "back to the Moon" narrative continued a decade later on Apollo 11's 50th anniversary, with discussions of returning to the Moon by 2024. With a still shot of the American flag planted on the Moon's surface, footage cuts to Orion, NASA's first design of a deep space crew capsule since the Apollo era. However, the reporter reminded viewers that it was years behind schedule and over budget (CBS, 7/21/19).

Additional narratives of funding, while less consequential, underscored what has been called the "greatest adventure story in human history" (CBS, 7/18/17). For instance, CBS covered a story in 2015 of an online campaign that raised $700,000 to preserve Neil Armstrong's space suit, far exceeding its goal of $500,000. The surplus amount was used to also restore Alan Shepard's Mercury suit, and both were sent to the Smithsonian Institution for display (CBS, 8/20/15). A few years later, covering another fundraiser, CBS returned to "one of the most famous rooms on Earth," Mission Control, where Mark Strassmann provided time-lapse footage of the room being dismantled and put back together with artifacts that matched vintage photographs of the famous room. "It was as though somebody had spit on sacred ground" (CBS, 6/29/19), said Mission Control Director Gene Kranz, who helped to raise $5 million to restore the room to its former glory. "This was the start of not only the space revolution, but the technology revolution within our nation. This is where it all began." Strassmann concluded the segment with, "It's back to the future at Mission Control" (CBS, 6/29/19). Though the room, equipment, and accoutrements had faded over time, this remembrance illustrated how important Mission Control was to Apollo-era space fans, who were willing to pay for its restoration, underscoring the value of the American accomplishment through its nostalgia.

HEROISM

Heroism, akin to the frontier narrative, has also been an important ingredient in television's recipe to appeal to American taste. NASA realized the potential long before missions flew, and the agency cultivated heroic images to build public support.[15] While NASA was more likely to promote Mercury astronauts as heroes, the space agency did spotlight the dangers and perils of space exploration, and the astronauts' intense training to overcome the risks.[16] NASA also realized its projects provided "good TV." From the beginning of CBS's Apollo coverage, astronauts were called "heroes," not necessarily because they explored the unknowns of space but because they illustrated the American moral ideal and personified the American Dream. Sevareid, for instance, described the Apollo 11 crew as individuals who grew up in "the folk image of the small Midwestern town" where "they had security" and "the leisure to prowl and to dream" (CBS, 7/15/69). Calling Neil Armstrong "*the* symbolic American" in the eyes of the world, Sevareid added that Armstrong typified "the all-American boy," "the kid next door":

> He has all his hair. He has frank blue eyes. His smile is a slightly shy half grin. And he has the inner strength to bear his country's pride to the rest of the world—strength he will need, not only for his country, but for himself and for his family. (CBS, 7/15/69)

Herbert Gans has noted that the celebration of small-town America—which Thomas Jefferson also revered—has been a celebrated news value, for it represents "the good life in America," restoring a sense of community and cohesiveness among citizens.[17] Here, Armstrong, from a small pastoral town in the Midwest, was the model American because he believed in himself and in his country. He and the other two astronauts were "individualists who function[ed] as cogs in a vast human machine," who "[saw] beauty in the machinery" and "[were] perfectionists seeking the outer limits of their strength and their talents" (CBS, 7/15/69). As men who would never question authority—"symbolic of the organization man, the cooperator" (CBS, 7/15/69)—Armstrong, Aldrin, and Collins looked like choirboys amid reports of protesters and rioters in American streets. Clergyman commented on the astronauts' "unselfishness and faith in God" at a ticker tape parade, then called them "American's latest heroes" (CBS, 1/10/69).

Even though each of Apollo 11's crew was referred to as "heroes," the media naturally devoted more time to Armstrong because he was the first to step foot on the Moon. In one segment, Eric Sevareid placed Armstrong in the company of two other heroes, Charles Lindbergh, the first American to fly an airplane solo across the Atlantic Ocean to Paris, and John Glenn, the first American to orbit Earth. Because CBS focused on the "never-before-done" aspect of a mission full of "unknowns" (such as not knowing if the LEM would sink as it touched down on the Moon), the astronauts' safety became a concern for many Americans. Like soldiers entering a hostile environment, so too did "the three young heroes of the space age" enter the black darkness of space, said reporter John Lawrence, and two of them faced a desolate and dangerous Moon while the third circled above in solitude (CBS, 8/13/69). Upon their return home, the astronauts were honored with televised parades and ceremonies—"heroes' welcome they so richly deserve" (CBS, 7/24/69)—not only in the United States, but as they traveled the world on their goodwill tours representing America. Again, NASA and CBS wanted the average American to aspire to be like the astronauts and to vicariously participate in the costly venture toward the skies. But in reality, both agency and network portrayed the "heroes" in a less than truthful manner. After Apollo ended, astronauts such as Buzz Aldrin publicly disputed the angelic media portrayals:

> Nearly all of it has us squarely on the side of God, country, and family. To read those accounts was to believe we were the most simon-pure guys that had ever been. This simply was not so. We all went to church when we could, but we also celebrated some pretty wild nights.[18]

Clearly, CBS played up its notion of what constituted an ideal American however false that vision might be.

The image was nonetheless short-lived. The role of the astronaut in telling the Apollo story plummeted after Apollo 11. Apollo 12 astronauts were treated more like "historical footnotes" than heroes (CBS, 11/13/69). Consistent with the goals after Apollo 11, the astronauts were instead portrayed as scientists on the Moon, there to accomplish a job and to teach the American public about scientific laws in space. This was evident in a segment that featured Apollo 15 astronaut David Scott demonstrating Galileo's law of falling objects and gravity fields by dropping a feather and a hammer simultaneously to illustrate (CBS, 8/2/71). But as the Apollo project neared its end, CBS was far more likely to point out blunders made by the astronauts than in earlier reports. For example, it showed Apollo 16 astronauts Charlie Duke dropping equipment and John Young tripping over and breaking wires while setting up scientific experiments on the Moon; both looked clumsy as they lost footing on several occasions (CBS, 4/21/72). A few days later, while Cronkite reported that part of a scientific experiment was lost when the astronauts "failed to flip a switch," CBS featured Bill Green, an engineer from North American Rockwell, who explained how *engineers* fixed another problem in the primary guidance system (CBS, 4/24/72). Just as in other romantic narratives, stories of heroism shifted from astronaut to engineer. Realizing the heroic icon was no longer needed, CBS followed NASA's lead in portraying the missions as scientific.

It would be many years before the spotlight would publicly fall upon the engineers. *Failure is Not an Option*'s narrator explained why the engineers have not been as well known as the astronauts.

> At any other time, these men would have had unremarkable careers. Now world events have put them at the center of a desperate struggle. Few will ever be

known to the public, but they will become the unsung heroes in the race to the Moon. (THC, 8/24/03)

The documentary went on to explain that astronauts often publicly thanked Mission Control for handling situations they themselves were unable to handle, while the engineers in turn gave God, NASA, or the American people credit. Solving problems required technical competency, and NASA's engineers had it, as will be discussed in detail in chapter 8. Thus, though not stated explicitly, NASA and contractor engineers became the new heroes in news reports, most likely to keep the faith in its technology despite a plethora of problems experienced by the astronauts.

This is not to say that Apollo's original helmsmen have been forgotten in contemporary representation, not by a long shot. The networks continue to depend on the familiar—the astronaut—to galvanize the American public, especially America's youth. For instance, CNN featured a segment—clearly aimed at younger viewers—called "Voices of the Millennium" in which it interviewed celebrities pertaining to a particular topic. The September 8, 1999, episode, entitled "Personal Heroes," was hosted by Apollo 13's Jim Lovell, who said his own hero was "a person like Charles Lindbergh, taking a flight across the Atlantic solo for the very first time." Similarly, a few weeks later, the topic of "Personal Heroes" was Apollo astronauts, who told young viewers who wanted to become "heroes" in the space program to get a good education and to study how things work (CNN, 9/29/99). Again, there was a focus on future exploration, and what better way to appeal to America than through its next generation of space enthusiasts?

As expected, CBS returned to the hero narrative of astronauts with news of their deaths. When Neil Armstrong passed away on August 25, 2012, CBS's Anthony Mason opened the broadcast with old footage of humankind's first Moonwalk and the "one small step for [a] man" video, calling Armstrong a hero who fulfilled Kennedy's goal. The broadcast went on to describe the astronaut's childhood in Ohio and his fascination with airplanes that "never abandoned him." Armstrong left Purdue University in 1950 when the Korean War broke out, flew 78 combat missions as a naval aviator, became a test pilot, and flew X-15 rocket planes to the edge of space. He exhibited his "trademark coolness" when Gemini 8 almost ended in disaster. Mason continued, "Armstrong had another brush with death when his training vehicle rolled to its side, and he was forced to eject just 200 feet from the ground. He then went back to the office to do paperwork" (CBS, 8/25/12). Mason then goes to interview footage with Armstrong: "That's true. There was work to be done." Mason: "You were almost killed." "But I wasn't," Armstrong replied with a smile (CBS, 8/25/12). Mason then chronicled the Apollo 11 mission with the standard footage of the Eagle landing, Armstrong walking on the

Moon, and Armstrong and Aldrin planting the American flag. "Despite all the initial fanfare after Apollo 11, Armstrong remained a humble and intensely private person" (CBS, 8/25/12). The segment ended with President Obama calling Armstrong "among the greatest of American heroes, not just of his time, but of all time" (CBS, 8/25/12). To highlight the hero status, two nights later, CBS reminded viewers that, although Armstrong was the first to walk on the Moon, "it almost didn't happen" as there was trouble with the alarm systems. Krantz gave credit to Armstrong's maneuvering and added, the "American who did it remained modest, always saying he was just part of the team" (CBS, 8/27/12). Once again, CBS spun the heroic tale of grit, skill, and humility, as did *First Man on the Moon*, stating Armstrong was just "true to himself" and what "you saw was what you got" (PBS, 12/3/14).

EMOTIONAL REWARDS

According to NASA, "There is an emotional, intangible dimension to the human presence in space."[19] Throughout its coverage and later in remembering Apollo, the networks continued to appeal to the emotions of Americans. These emotional rewards, taken in many cases from NASA's playbook verbatim, included challenge, inspiration/human spirit, adventure/drama, spirituality, peace, and satisfying curiosity. Not only are these emotional rewards analogous with the "American spirit," but they also provide all the ingredients needed for "good television." The networks, then, could increase audiences while simultaneously reporting space exploration stories. As will be evident, it was through these rewards that Apollo became a ritualized display of Americanism.

Challenge

The American will to overcome challenges was discussed in the Nationalism section in terms of meeting Kennedy's goal. But the rhetoric of challenge was also presented by the networks as a romantic, symbolic way of accomplishing *any* goal, and Apollo was its prime example. For Flight Director Gene Kranz, Apollo demonstrated "What America can dare, America can do" (THC, 7/21/04). For other NASA engineers, such as Bob Carlton, not only was Apollo a challenge, but it proved that anything is possible. "When I was a kid and someone wanted to express something so impossible, [something] beyond comprehension, they'd say, 'You can't any more do that than you can go to the Moon.' And there's the Moon," he said as a close-up photograph of the lunar surface was shown (THC, 8/24/03). The Moon itself continued to be described symbolically by CBS. Charles Osgood, for example, called it the "romantic symbol of all that is unattainable." He continued, "We would say, 'You're asking for the Moon,' meaning it was impossible. Then, 10 summers ago, we did the impossible; we went there" (CBS, 7/19/79). A CBS newscast later devoted almost seven of its 30 minutes to the 20th anniversary of the Moon landing on July 20, 1989. Calling Apollo "the greatest achievement in the history of the U.S. space program," then- President George H. W. Bush capitalized on the historical date to set ambitious goals for America's future in space, including a trip back to the Moon with a permanent settlement there, and to Mars, declaring that space exploration is still "the inescapable challenge" for the United States (CBS, 7/20/89). Similarly, Armstrong's speech on CNN's 25th anniversary of Apollo also focused on the romantic theme of challenge:

> Today we have a group of students, among America's best. To you we say we've only completed a beginning. We leave you much that is undone. There are great ideas undiscovered, breakthroughs available to those who can remove one of the truth's protective layers. There are places to go beyond belief. Those challenges are yours in many fields, not the least of which is space, because there lies human destiny. (CNN, 7/20/94)

The same sappy rhetoric of school graduation speeches—promising graduates that opportunities and challenges aplenty await them—continued to animate reports of Apollo. Little had changed in terms of the rhetoric to persuade viewers to buy into the space program. Because of its success, Apollo was often recalled when current space projects such as Hubble, Challenger, or Columbia failed, calling the Moon era the "glory days" of NASA because there was a *goal*: get to the Moon before the decade was out. Thus, while the late 1960s and early 1970s were portrayed as halcyon years for NASA, in reality, Apollo was not without problems, especially with missions 1 and 13. Networks, of course, were far less likely to be critical of those mishaps.

Inspiration and Human Spirit

Other narratives romanticized the Apollo missions as inspirational, driven by "the human spirit" (CBS, 12/24/68). Here, the goal was to inspire Americans to reach for their own stars as the United States aimed for the Moon, a "dream [of man] since the beginning of time," according to Cronkite (CBS, 11/20/68). Networks exercised these narratives in three ways. One was to weave into the Apollo tale additional inspirational stories and link their commonalities. For instance, stepping back in time and invoking Archimedes's discovery of the calculus involved in the movements of the Earth and Moon, Sevareid described how Apollo 11 astronauts danced on the Moon just as

Archimedes danced in the streets of Syracuse, each feeling "young at heart" (CBS, 7/21/69) at discovering the mysteries of the heavens and Earth. This youthful feeling spawned by the astronauts was contagious:

> Not everybody felt born again exactly, but everyone must have felt young again, if just for a moment. It was the same 42 years ago when the boy Lindbergh touched down in Paris. F. Scott Fitzgerald wrote, "People sat down their glasses in country clubs and speak-easies and thought of their own best dreams." (CBS, 7/21/69)

The power of the astronauts was to evoke the roots of Americanism—people realizing "their own best dreams." Speaking of Lindbergh had created a pattern of association: U.S. technological success linked to model citizenship. President Nixon also took part in this romantic rhetoric when he said the Apollo 11 mission lifted the "spirits of the American people and the world" (CBS, 7/15/69). Similarly, Nixon, in a televised speech praising the astronauts on the day they returned to Earth, told the astronauts that, because of them, "all of us in government, all of us in America . . . can do our jobs a little better, we can reach for the stars, just as you have reached so far for the stars" (CBS, 7/24/69). Again, this theme highlighted a bright spot amid social discord—an argument for space exploration that would resurface years later as a way to bring people closer together in the midst of world strife. As an Apollo 8 viewer wrote in a letter to NASA, "Thank you. You saved 1968."[20] Even 50 years later, for the Apollo 11 anniversary, CBS's Reena Ninan waxed poetic, "50 years ago [Armstrong] became the first man to walk on the Moon. The skies have never looked the same. Apollo 11 was a defining human achievement and made people dream that space as within everyone's reach" (7/20/19). Similarly, Norah O'Donnell captured the feeling of inspiration as she recalled the historic flight.

> If the sky is clear tonight, you will see a full Moon shining brightly, as if the Heavens had planned this fitting tribute to the greatest scientific achievement of all time, one that stands as an inspiration to every other seemingly impossible challenge. (CBS, 7/16/19)

A second and significant method by which networks evoked human spirit was by featuring actual clips of ordinary citizens who were inspired by the Apollo story. Perhaps none of the narratives for Apollo 15 was more inspirational to viewers, especially young viewers, than the story told by C. J. Underwood, a reporter from CBS-affiliate WBTV in Charlotte, North Carolina (CBS, 7/23/71). His nearly five-minute segment three days before the launch featured 16-year-old Brad Perry, "a fourth American astronaut" of Albemarle, NC, who built an *exact* replica of the Apollo 15 command

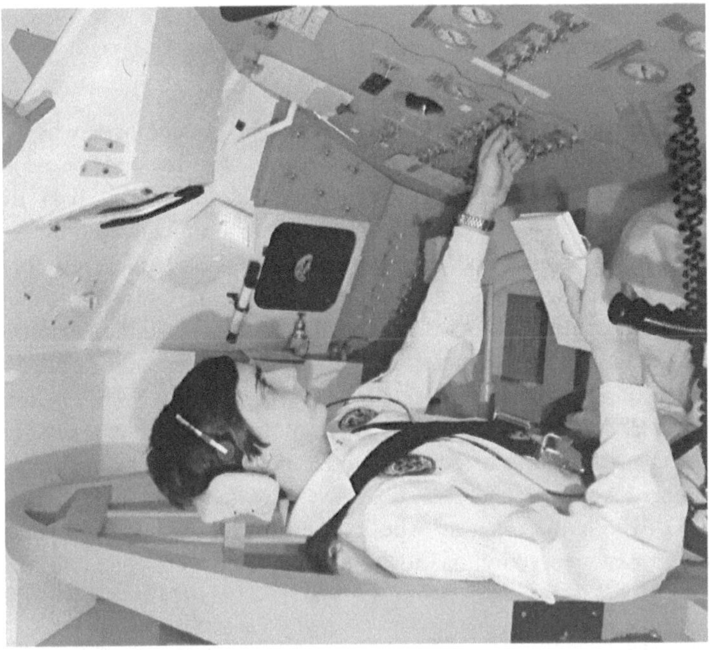

Figure 6.3 Brad Perry in his Apollo 16 Simulation, 1971. *Source:* Courtesy of Brad Perry.

module in the basement of his home. The segment opened with Perry in his bedroom adorned with NASA Moon posters on the walls and model rockets atop the furniture. What once seemed like "mission impossible" was now "an achievement incredible," said Underwood, noting that it took two years and a cost of more than $2,000 to build the module, paid for by Perry's paper route. With the help of his parents and later from NASA, which gave Perry technical data including the flight plan, Perry would spend the next 12 days in the module, living just as the astronauts do in space, in a spacesuit and all (figure 6.3). Asked if he wanted to become an astronaut, Perry said that he would go to the Moon if given the chance. In marveling at Perry's scientific activities while in the module, Underwood ended the segment, "The simulated flight he is taking says a great deal about why man indeed has reached the Moon," asserting that, if you work hard and save your money, you can become anything you want to become. Americans are naturally ingenious.

Additionally, this segment cast Perry as a role model to get children interested in science and space exploration. As Perry stated years later,

> As a boy, I saw everything NASA did as a success. I thought Apollo 13 was exciting and scary and challenging. I never doubted for a minute that they would

> be safe. It reinforced my belief that anything was possible when people worked together as a team.[21]

Perry, 50 years of age and working for NASA as an engineer when I interviewed him, said that he became interested in space exploration as a result of watching the television coverage of Apollos 7 and 8 (figure 6.4). Reading about Apollo in *National Geographic* and *Life* also influenced him. Asked if his interview with Underwood had been scripted, Perry answered no. "They would tell me what we would talk about. They gave me a few of the questions in advance so I could think of what to say, but nobody told me what to say."[22] CBS and NASA shared a common goal: cultivate an interest in space exploration. The Apollo program was in its planning stages when NASA Administrator James Webb wrote, "Among the major motivations of the space program [are] . . . the stimulating effects on all segments of American society, particularly the young."[23] Echoing NASA's sentiments, years later, Senator Bill Nelson (D-FL) claimed that an Apollo-like program would "rejuvenate and invigorate a new generation of kids to be interested in science and math and

Figure 6.4 Brad Perry in Front of His Apollo Command Module Mock-up on Display in His Hometown of Albemarle, North Carolina for Apollo 11's 50-Year Anniversary, 2019. ***Source:*** **Courtesy of Brad Perry.**

engineering just like we did back in the 60s with Apollo;" otherwise, "we become a second-rate nation" (CNN, 1/14/04).

As expected, networks continued to find ways to illustrate how stories inspired children who would one day become the next generation of space explorers, or stories recounting youthful Apollo astronauts. *First Man on the Moon* provided an anecdote of how Neil Armstrong, who obtained a pilot's license before a driver's license, built a wind tunnel in the basement of his childhood home that shook when he tested his model airplanes to figure out how to make them fly higher and faster (PBS, 12/3/14). To further illustrate, for Apollo 11's 50th year commemoration, CNN featured Michael Massimino, who, as a six-year-old child, aspired to become an astronaut after watching Apollo 11 astronauts land on the Moon. The story opened with a narrator stating,

> The fact is, that moment still inspires us, and it still sends a chill up the spine of anyone who was alive and watching that grainy footage live on that hot July night 50 years ago, either from Mission Control or their living rooms.

Massimino added, "It was magical for me," as the interviewer responded, "You are proof that it pays off to go after that dream and never give up." Viewers were then told that Massimino was rejected three times by NASA before he became an astronaut (CNN, 7/20/19).

Similarly, for its reporting on Artemis 1, the first mission to the Moon in 50 years, CNN featured the story of African American NASA astronaut Victor Glover. "For me, Artemis I is exciting, but it's really a steppingstone, a milestone to getting humans back in the vicinity of the Moon, and that is awesome!" (CNN, 8/28/22). Reporter Christina Fisher added, "Victor Glover is one of the more than 40 astronauts in the running to fly on Artemis II and Artemis III, which will land the first woman and the first person of color on the Moon," followed by Glover pointing out that, "We explore *for* all people, but now we can actually say we explore *with* all people" (CNN, 8/28/22). The television screen then showed three men and five women, all part of the new astronaut corps, wearing astronaut uniforms. Every anniversary became an opportunity to appeal to those who were traditionally excluded from yesterday's astronaut club.

Finally, networks also inspired viewers by showing images of the now-famous Earthrise photograph from Apollo 8 and the Earth View photograph from Apollo 17. In remembering the tenth anniversary of Apollo 8, for instance, Cronkite featured Earthrise (figure 6.5) as he ended the newscast with, "We don't go to the Moon anymore, because 10 years ago we traveled to the Moon and discovered the Earth" (CBS, 12/25/78). Interestingly, the Earthrise icon even became the *CBS Evening News*' backboard, shown

Figure 6.5 Apollo 8's Earthrise. *Source:* Courtesy of NASA.

between newscasts and in segues to commercials in the late 1960s and early 1970s. Amid criticisms that the space program no longer had a goal or purpose as it had in the Apollo days, for Apollo 11's tenth anniversary, CBS reminded viewers that Apollo "lifted the spirits of people throughout the world," as Nixon said ten years prior in 1969 (CBS, 7/15/79).

On the other hand, the Earth View photograph (figure 6.6) took the emphasis from the Moon and placed it on the Earth. It spawned a worldwide environmental movement to inspire people to take better care of the Earth, with groups in the 1970s adopting the image to champion their causes. Whereas the Earthrise photograph "helped extend America's Manifest Destiny into the ultimate wilderness of space," the Earth View photo illustrated the fragility of Earth alone in the vastness of space.[24] Apollo 8's Bill Anders remembered:

> We came all this way to the Moon, and yet the most significant thing we're seeing is our own home planet, the Earth. . . . All of the views of the Earth from the Moon have left the human race and its political leaders and its environmental leaders and its citizenry realizing we are all jammed together on one really kind

> of dinky little planet, and we better treat it and ourselves better or we're not going to be here very long. (PBS, 7/13/99)

Media representations of two of Apollo's most famous photographs invoked a geopolitical message as well, making us "realize just what you have back there on Earth" (PBS, 1/25/78). Calling the Earth "a grand oasis in the vastness of space," *One Small Step*'s narrator also reminded us that, although the Apollo program was created out of military or political motivation, the results were longer lasting than the competition with the Soviets had been (PBS, 1/25/78). The Earth View photograph was pictured as journalist Richard Lewis added,

> The journeys to the Moon gave us a new perspective about the Earth and enabled us to visualize ourselves as living on a planet afloat in space, a visualization which is particular to the 20th Century, in which I believe had never existed before in human consciousness. (PBS, 1/25/78)

By now, the two photographs have become etched in the public consciousness. Depending on the familiar is consistent with what William Cohn has

Figure 6.6 Apollo 17's Earth View. *Source:* Courtesy of NASA.

argued: television in general, and documentaries in particular, cater to the tastes of mass audiences, choosing familiar images and notions of American history that will fulfill viewers' expectations and draw them toward positive consensus.[25] For what more could NASA and the networks ask?

Adventure and Drama

Like NASA, which claimed humans have an innate will to be venturesome, CBS framed the missions as exciting from the beginning of its Apollo coverage. Cronkite described the flight of Apollo 7 as attracting a "morbid curiosity" among the spectators, yet providing a "vicarious thrill" (CBS, 10/8/68) to every person who imagined being aboard the tiny capsule that left Earth orbit. But CBS, as a television network whose goal was to increase viewership, dramatized all it could. Akin to the rivalry between the United States and the Soviet Union that added a dramatic flair to the coverage of Apollo missions, when the Apollo 10 astronauts performed technological maneuvers for the first time, and when the spacecraft flew around the dark side of the Moon sans radio contact with Mission Control on Earth, reporter Bruce Morton called the events "heart-stopping" and "suspenseful" (CBS, 5/23/69). Called "the true climactic of intense drama," the Apollo 11 mission would be "perhaps the longest drawn-out performance ever staged" according to Sevareid (CBS, 7/17/69). As the world would soon learn, the real spectacle was yet to come with Apollo 13, the "mysterious accident," a "drama in space" (CBS, 4/14/70). "History's first celestial cliff-hanger" (CBS, 4/17/70), a "chilling adventure in space" was also a "crisis," yet an "excitement" (CBS, 4/14/70). CBS noted that Americans were not the only ones who found the mission to be riveting; the accident had captured the largest world audience of all time (CBS, 4/17/70).

The dramatic clips also served to evoke nostalgia, as the Apollo 11 crew said they had hoped that a new generation would recapture the excitement that was present 20 years prior (CBS, 7/20/89). And what better medium to illustrate drama than television? Tellingly, THC depended on CBS footage—not NASA's—to tell the Apollo story and to create the original ambiance of the time to further dramatize the events. By incorporating vintage CBS footage of Cronkite, THC could re-dramatize the Moon shots. *Triumph and Tragedy*, for instance, mentioned the scare just before the Eagle had landed on the Moon when Armstrong was forced to make a manual landing because the fuel supply of the LEM was dangerously low. Once THC confirmed that the astronauts had landed safely, it reverted to the familiar: Cronkite giggling, "We're home. Man on the Moon. Oh, geez! Oh, boy," while Schirra was pictured wiping a tear from his eye (THC, 3/10/99). As the *Tech Effect: Apollo 11* narrator approached the end of the story, he mentioned that, when it was time to leave the Moon, the astronauts did not know if the engine would start or if they would have enough fuel to blast off from the lunar surface. For the first time, viewers learned that the situation was so tense that NASA suggested to the White House that President Nixon prepare a speech for the American public in case the astronauts were stranded on the Moon, in which case they would have perished within days (THC, 6/20/04).

THC's strategy of recycling appealed to later audiences. It paired known Apollo stories with others never before told, usually pertaining to the engineers. For instance, *Days that Shook the World: The Wright Brothers and the Moon Landing* and *Failure is Not an Option* both told the story of Steve Bales of Mission Control. Here, THC coupled an unfamiliar story of the Apollo program with the familiar to dramatize. In *Days That Shook the World: Wright Brothers/Moon Landing*, as Bales was pictured walking down a hallway at NASA, the narrator indicated, "He doesn't know it, but the whole Apollo [11] mission rests on *his* shoulders" (THC, 11/29/03). It was Bales's decision whether the crew should land or abort the mission. According to this broadcast, communications were almost completely lost between the astronauts and Mission Control, and the Eagle was descending at a rate far too fast—so fast that Armstrong's heart rate jumped from 77 beats per minute (BPM) to 156 BPM. Members of Mission Control, said the narrator, also feared the spacecraft would crash into the lunar surface. When the computers—which were the pinnacle of 1969 technology with 74 kb of memory, less than a modern-day cellular telephone—sent signals warning Mission Control engineers to abort the mission, 26-year-old Bales had a split second to decide whether to land the Eagle (THC, 11/29/03). Fearful that the lunar lander might crash, Bales gave the go-ahead to land nevertheless and later would receive a Presidential Medal of Honor. Calling Apollo "the greatest adventure story in human history" (CBS, 7/18/17), these unfamiliar details refreshed historical events for another generation of viewers.

Spirituality

From Apollo's inception, NASA spoke of "the spiritual awakening which accompanies the pursuit of a new, vast, and enormously challenging goal."[26] With most of the astronaut corps professing faith in God, it is no surprise that the networks depended on spirituality to appeal to Middle America. They accomplished this in several ways. One was to employ Biblical or creationist metaphors. On Christmas Eve 1968, Astronaut Frank Borman read a prayer as *CBS Evening News* showed Earthrise pictures. After he read the prayer, he told audiences nearly 250,000 miles away that he was sorry to have missed church on Christmas Eve. The following day CBS broadcast audio and visual clips of the "unexpected message" from astronauts reading Bible verses from the book of Genesis, with Bill Anders commenting,

> We are now approaching lunar sunrise, and for all the people back on Earth, the crew of Apollo 8 has a message that we would like to send to you: "*In the beginning God created the heaven and the earth. And the earth was without form, and void; and darkness was upon the face of the deep . . .*"

Each of the crew read from the scripture, and Borman finished by saying, "'. . . *and God saw that it was good.*' And from the crew of Apollo 8, we close with good night, good luck, a Merry Christmas, and God bless all of you—all of you on the good Earth" (CBS, 12/25/68). Cronkite ended with "and that's the way it is, this Christmas Day, 1968" (CBS, 12/25/68). Years later, engineers such as Jerry Bostick remembered on *Failure is Not an Option*, "To hear that coming from Americans circling the Moon on Christmas Eve was a very emotional moment, a very proud moment" (THC, 8/24/03). The narrative spoke to the pride NASA personnel had in God and country. It worked. In the dawn of the space age, science and technology representatives still revered the ancient Word.

NASA denied planning Apollo 8 to fall at Christmastime and insisted that engineers planned the missions according to windows of opportunity such as clear weather, the position of the Moon, and contractor schedules concerning the delivery of Apollo machinery. Neither NASA nor CBS planned the biblical reading; the idea came from a friend of Frank Borman who worked for the U.S. Bureau of the Budget.[27] Even so, Madalyn Murray O'Hair, whose Supreme Court case took prayer out of public schools, later filed a lawsuit against NASA in an Austin, Texas federal court because the Apollo 8 astronauts read Bible scripture "while on duty." She asked for an injunction to prevent future astronauts, while on official government time, from engaging in religious practices, Cronkite reported to audiences (CBS, 8/6/69). The announcement of the suit created a massive wave of support for the mission, and within a year, NASA received over 2.5 million letters and petitions in favor of the astronauts' First Amendment right to publicly exercise their free expression of religion.[28] The Court agreed and later dismissed the case (CBS, 3/8/71). Americans clearly supported the sacred references. Even if NASA had not planned the spiritual aspect of the mission, the astronauts did. The stunt would rehearse the "goodness" of the astronauts.

Spirituality also crept into the news, whether in celebrations of safe journeys or to illustrate America's faith in times of social unrest. For the former, at parades and receptions, clergy praised the astronauts for "[giving] us faith when we really needed it," and said that their "unselfishness and faith in

God" illustrated that we as a nation "should go higher" (CBS, 1/10/69). For the latter, broadcasts featured footage of PresidentNixon speaking at a prayer breakfast with Reverend Billy Graham as a special guest. There, Nixon told audiences that the White House received many letters from people who claimed they were praying for the astronauts and the United States. Through prayer, he said, the United States could meet its many challenges sustained by faith (CBS, 1/30/69). The ritual of prayer spoke to the commonality of Americans.

If displays of spirituality underscored America's beliefs, they also served to appease its doubts. The networks accomplished this in two ways. One was by employing religious metaphors for the Apollo experience itself. For instance, adding to the sense of history being made was what Sevareid called a "reverential sensation" (CBS, 7/16/69) associated with Apollo 11 when Cronkite reported that man was on his way "to the heavens" (CBS, 7/16/69). Sevareid told audiences, "This really *is* a religious experience. What you watch is a biblical scene. The ground literally trembles. The air hits you in the face. And all that flame that comes out of the motors, it's a whole ocean of flames" (CBS, 7/16/69). He seemed to conjure up angels rising from blazes of fire. Because technology enabled man to "finally talk to God,"[29] science fiction writer Ray Bradbury predicted, "By the end of the century, our churches will be full again. [Apollo 11] is redesigning mankind back in the direction of God again" (CBS, 7/21/69). Americans distraught that prayer had been taken out of public schools six years earlier may have been encouraged to take heart.

The second way in which CBS appeased doubts was to demonstrate faith in technology and the technicians who built the "chariots for Apollo."[30] Spiritual metaphors enhanced the Saturn V rocket and other Apollo 11 technology. For example, former President Lyndon B. Johnson called the Saturn V "that great power" (CBS, 7/16/69), but Sevareid went much further.

> The great spire stands there like a modernized gothic cathedral. And like the medieval cathedral, this mobile temple was built by a special breed of anonymous artisans, thousands of them, who will then move on from site to site as high priests of this new religion of space instructor. (CBS, 7/14/69)

The "artisans" or engineers who created this complexity became "high priests" in "the new religion." In his string of metaphors, the Kennedy Space Center became "holy ground," a "biblical scene" (CBS, 7/16/69), what Thomas Pynchon later called the "Holy Center" in *Gravity's Rainbow*.[31] In addition to marshalling public opinion in support of the project, touting the spiritual might have been CBS's way of easing viewers' fears of heaven-bound technology. The Creator, "the fourth astronaut up there"

(CBS, 4/14/70), was ultimately in charge and would bring the crew home safely. As Apollo 8's Borman said, "Let's face it. God must have been looking after us or we wouldn't have made it, so it was a very, very, very religious experience for me" (PBS, 7/13/99). Such portrayals were reminiscent of what NASA's Wernher von Braun had announced before Congress a few years prior: "There is a force greater than the thrust of [astronauts'] rocket ships, a spirit greater than the cold logic of their competitors, a power greater than their own nation."[32] The ritual of prayer and faith in God, then, appealed to the audience—narratives they not only understood, but more importantly, endorsed.

Such symbolism ensued in remembrances, but to a lesser extent. When Apollo 11's Neil Armstrong died, CBS reported from the Washington National Cathedral where his funeral was held. In its "Space Window" that depicts the Apollo 11 mission, there is a piece of Moon rock embedded in a stained glass, a gift from NASA. According to the reporter, Armstrong "showed us the pathway to the stars," while Cernan, who was one of many who spoke, said, "You can now finally put out your hand and touch the face of God" (CBS, 9/13/12). This, of course, echoed the aviator's prayer from the poem by John Gillespie Magee, titled "High Flight," that President Ronald Reagan used to honor the lives of seven astronauts who perished when the Space Shuttle Challenger exploded on January 28, 1986.

Peace

Although the first Apollo missions seemed to pit the United States against the Soviets, Apollo 11 was portrayed as bringing the world's citizens closer together. Peace was a recurring theme in *CBS Evening News*. Harry Reasoner even suggested a U.S.-Soviet joint mission to Mars to make "harmony possible between the two countries" (CBS, 7/10/69), while on CBS Ray Bradbury predicted, "We really are three billion lonely people on a small world and I think [the Apollo 11 mission] is going to draw us much closer together" (CBS, 7/21/69). The clichés multiplied into a critical mass. The utility of cooperation and the rhetoric of world peace were political sound bites in the Nixon White House. CBS provided coverage of Nixon speaking to an audience of 2,000 foreign students from 60 countries gathered on the White House lawn about the success of Apollo 11, "in terms of what we can do on this Earth."

> I hope that when the next great venture into space takes place that it will be one in which Americans will be joined by representatives of other countries so that we can go to new worlds together. . . . I thought one of the, shall we say, rather sad things about that great day on Monday, when man first stepped on the Moon, was that, while most of the peoples of the world saw it on television,

> or participated in it on television or radio, that there was approximately a half the world that did not see it—the world of Communist China, the world of the Soviet Union. I thought of how sad that was. Sad not in terms of East-West conflict, because this is no time to discuss that, but sad in terms of the people involved. And I want the time to come when the Chinese people and the Russian people and all the peoples of this world can walk together and talk together as you walk together and talk together. (CBS, 7/22/69)

The next day, Nixon told reporters of an upcoming world trip he would take after attending a welcome-home dinner for the crew of Apollo 11. His mission, he said, would be to spread international brotherhood around the world. He then added, "The United States will show that we won't abdicate world leadership to Moscow or Peking" (CBS, 7/23/69), a marked contradiction from the statement a day before.

Politicians were not the only people who publicly tried to unpack the meaning of Apollo. Months after the Apollo 11 landing and a couple of weeks before Aldrin, Armstrong, and Collins participated in a 38-day goodwill tour in 24 nations, Aldrin announced to Congress, "As the Moon shines impartially on all those looking up from our spinning Earth, so do we hope that the benefits of space exploration will be spread equally to the harmonizing influence to all mankind" (CBS, 9/16/69). Some might have wondered whether the "harmonizing influence to all mankind" was simply an attempt to persuade Congress of the political benefits of space exploration to secure funding for future exploration.

Naturally, most of the references to Apollo's peaceful potential were ideological. In their cramped quarters, the three Apollo 8 astronauts lived peacefully for six days, sending the message that "we should all learn how to get along better" (CBS, 1/10/69). On cue, CBS featured New York Mayor John Lindsay saying that we should learn from the Apollo 8 mission how to live together: "How simple it is when man is matched against perils of an unknown and physically hostile universe" (CBS, 1/10/69). Before the Apollo 11 mission, Cronkite insisted that the purpose was to learn about Earth and therefore ourselves (CBS, 7/7/69). After Apollo 11, the focus became the lack of political boundaries. As Armstrong publicly noted in Mexico on a goodwill tour: "[From space] we could see no boundaries between countries; we could only see our hemisphere—its continents joined in a manner fitting to the way our people have been joined over the years" (CBS, 9/29/69). Similarly, on a trip to West Berlin's "Wall," Mike Collins observed, "It is not possible to build walls in space" (CBS, 10/13/69). At least Apollo no longer implied conquest. It now meant cooperation and friendship.

Peaceful intent attached itself to losses as well. However crippled the Apollo 13 mission had become, it was also a "creator of human solidarity"

(CBS, 4/14/70), "remembered because just for one brief moment, all of the people of the world were united" (CBS, 5/5/70). Amid tiresome Vietnam War reports, world peace seemed to be a palpable reason to continue exploration. It continued. *To the Moon* featured footage of Apollo 17's Cernan as he climbed up the ladder to the lunar module Challenger to head back to Earth and said, "We now leave as we once came, and God willing we shall return with peace and hope for all mankind. And may America's challenge of today forge man's destiny of tomorrow" (PBS, 7/13/99). It seemed that the American technology of landing on the Moon was not enough—the networks wanted to give the U.S. credit for providing world peace as well.

Satisfying Curiosity

In a final romantic narrative, NASA capitalized on humankind's innate need to satisfy curiosity. In 1963, the agency claimed,

> For centuries man has looked to the skies and sought to break the chains which shackled him to a single planet in the vast universe. This aspiration has stemmed not only from his curiosity but also from man's fundamental thirst for knowledge.[33]

CBS again reiterated NASA's formula: it appealed to America's insatiable quest for knowledge, and the Moon, "a source for fantasy" that had long intrigued humankind (CBS, 7/29/69), was now its object. For instance, Charles Kuralt offered an eight-minute history of the Moon as a background piece to the Apollo story (CBS, 12/23/68). Opening the segment with a vintage recording of "My Sweetheart's the Man in the Moon," Kuralt explained that, while the Moon might be the target of the American and Russian space programs, it also evoked emotional fascination among the public. The segment rehearsed the legend of the new Moon's fertility, told of why some cultures thought of the Moon as divine, noted stories of the "dark side of the Moon," and dwelled on the origins of terms such as lunacy. Films of bygone days jostled against tales by Jules Verne. Kuralt ended with,

> We know the real Moon, front, back, warts and all. We know what she weighs and how she wears her hair. All that remains now is to go. . . . Our generation is different than any other in the billion years since life on Earth. Yet it is possible, in that splendid moment when we gain the Moon, it is just possible that our first emotion will be a shocking sense of loss. (CBS, 12/23/68)

The segment reminded audiences that technology could dispel mystery and diminish cherished traditions. No longer composed of "dirty beach sand" and "plaster of Paris" as the Apollo 8 astronauts had described it

(CBS, 12/24/68), the Moon, with its "precious" and "priceless stones" was a "geologist's Moon" (CBS, 7/14/69). "The old familiar Moon of lovers, poets, astronomers, songwriters, and most of the rest of us," said Cronkite, seemed faded (CBS, 7/8/69). The Apollo missions would forever change the Moon because humankind would know for certain what it looked like and from what it was made. Once the Moon's "warts and all" had been discovered, many began to voice opposition to continuing the trips. Even before the climax of Apollo 11, coverage of Apollo 10 revealed that space exploration had become routine because the missions were operating so smoothly. It was no wonder that the public began to lose interest. Curiosity seemed sated.

Perceptions of routine doubtless contributed to arguments that space funds should be used instead to solve problems on Earth now that the United States had satisfied its curiosity about the Moon. During a televised service to dedicate the Goddard Memorial Library in honor of the American rocket pioneer Robert Goddard, Senator Edward Kennedy (D-MA) said there was no need to be involved in a space race with the Soviets and that the space program must be in harmony with national priorities in solving problems at home (CBS, 5/19/69). To illustrate, in his commentary, Sevareid explained,

> We need to concentrate on inner space problems and ask ourselves why we can fly to the Moon but cannot solve the problem of the city's garbage or the mysteries of the 5:00 traffic jam. . . . It is not going to do us much good, to understand the nature of the physical universe, until we unravel the nature of man. (CBS, 5/26/69)

Thus, the metaphor of "social healing" was resurrected. Now that the goal was in sight, for some that meant moving on to other, more worthy projects. It was an argument that continued until the last mission. Familiarity may not have bred contempt, but it certainly boosted doubt.

Perhaps no other CBS news segment illustrated the argument for replacing space travel with social healing more than Harry Reasoner's report on the Southern Christian Leadership Conference (SCLC) demonstration at Kennedy Space Center "to remind Americans if [they] spend millions on space, then they could afford to feed the poor" (CBS, 7/14/69). CBS covered the "poor people's campaign" after SCLC's president, Dr. Ralph Abernathy, asked NASA Administrator Thomas Paine to allow several poor families—mostly African American—to be admitted to the V.I.P. area to view the launch in person (CBS, 7/15/69). Although Paine granted Abernathy's request, he "told Abernathy that if he could wipe out poverty by not sending Apollo 11, he'd not push the button" (CBS, 7/15/69). On the day of the Apollo 11 launch, CBS News again featured SCLC and Abernathy, who stated:

> There's a great deal of joy and pride. For that particular moment and second, I forgot that we have so many hungry people in the United States of America. But now I remember that we will have to go back to business as usual in trying to really launch a program that will move off on schedule and with the speed and rapidity as did this marvelous and magnificent rocket. I was one of the proudest Americans as I stood on this soil, on this spot. I think it's really holy ground and it will be even more holy once we feed the hungry and care for the sick and provide for those who do not have houses. (CBS, 7/16/69)

Next, CBS showed the group standing together singing, "We Shall Overcome," the anthem of the Civil Rights Movement of the 1960s,[34] while beeps from outer space played under the audio of the group singing. CBS then cut to Walter Cronkite, who said, "Whether or not we shall indeed overcome on Earth, tonight man is overcoming the skies" (CBS, 7/16/69). Unable to offer solutions to social situations, CBS seized on the much easier strategy of pretending to understand the technological problems pertaining to space exploration. Tensions between social problems and technology surfaced in Sevareid's commentaries on the day after Armstrong descended the Eagle's ladder onto the Moon:

> Any number of horseback philosophers finds it most odd that such a clean, beautiful, successful act of human cooperation could be born in the middle of so much acrimony, failure, and violence. But there's never been a logical nation or a logical epoch. Alfred North Whitehead said, in fact, that every period of great human creativity has been accompanied by or immediately preceded by much mass violence. It means apparently, that there just happens to be times when human drives come to a boil, then you get arson in the streets and flights to the Moon. (CBS, 7/21/69)

Thus, despite the technology, said the "horseback" commentator, humans will act according to their nature. It might be man's desire to create, but it is also his will to destroy, especially during wartime.

Realigned priorities also appeared in reports of funding the Vietnam War. On the day after the lunar landing, CBS's Harry Reasoner, with the Apollo 11 mission patch projected behind him in the background, reported that the astronauts safely landed on the Moon, and then introduced a report that described a typical day for "other Americans" in a dangerous part of the Earth: South Vietnam (CBS, 7/21/69). As Richard Threlkeld reported, "Today was no national holiday here. Tired, dirty American GIs came back from ambush patrols to learn that two of their countrymen had landed on the Moon. But what was more important—[the GIs] had survived another night on Earth" (CBS, 7/21/69). When a reporter asked a soldier what he was doing when the astronauts landed on the Moon, the soldier replied, "I just got out of an ambush" (CBS, 7/21/69). This, unfortunately, was one of only a handful of reports that drew a connection between Apollo and the Vietnam War, two events that helped define 1960s America. To have offered further analysis of the conflicting events would have harmed the cultural reality that CBS had worked so hard to create and maintain.

CBS did, however, highlight the mounting costs of the missions in reports after Apollo 11, in addition to focusing on challenges that astronauts faced while on the Moon. Although the entire program cost about a seventh as much as the Vietnam War,[35] CBS News reminded viewers that Apollo 14 totaled $400 million, $25 million more than Apollo 13 (CBS, 1/30/71). Knowing that the flights to the Moon were exorbitant, watching the astronauts play golf on the Moon and conduct personal experiments in space might have led audiences to question the need for yet another costly mission, especially those riddled with so many problems, such as the near-tragic Apollo 13 still fresh in their minds. Additionally, in covering Apollo 15, Bill Stout reported that the command module, the LEM, and the lunar rover cost more than $10 million apiece, "for a total cruising on the Moon of 120 miles" at eight miles per hour (CBS, 3/12/71). However, immediately after divulging the costs, Cronkite reiterated the scientific worth of the missions, which were increasingly a method by which to advance the program. Years later, *To the Moon's* narrator offered the insight that, if man was to return to the Moon, it would not be out of political necessity, but rather because of scientific curiosity, and would be funded not by the government, but rather by private enterprise (PBS, 7/13/99). Curiosity, then, became less romantic and more pragmatic in the networks' quest to retain viewers, as will be the focus of chapter 7.

NOTES

1. Ruth Schwartz Cowan, *A Social History of American Technology* (New York: Oxford University Press, 1997), 210.
2. Ibid., 213.
3. Mark E. Byrnes, *Politics and Space: Image Making by NASA* (Westport, CT: Praeger Publishers, 1994), 47.
4. Ibid., 48–49.
5. Howard McCurdy, *Space and the American Imagination* (Washington, DC: Smithsonian Institution Press, 1997), 143.
6. Roger D. Launius, "Perceptions of Apollo: Myth, Nostalgia, Memory, or All of the Above?" *Space Policy* 21 (2005): 131.
7. Bill C. Malone, *Country Music, U.S.A.* (Austin, TX: University of Texas Press, 1985), 317.
8. Bill C. Malone, *Don't Get above Your Raisin': Country Music and the Southern Working Class* (Urbana, IL: University of Illinois Press, 2002), 42.

9. Michael Gray, Country Music Hall of Fame, E-mail correspondence with author, November 4, 2004.

10. Adam Chaikin on THC's *Modern Marvels: Apollo 11*. Originally aired July 21, 2004.

11. Michael Schudson, *Watergate in American Memory: What we Remember, Forget, and Reconstruct about the Past* (New York: Basic Books, 1992), 153.

12. Vernon Van Dyke, *Pride and Power: The Rationale of the Space Program* (Urbana, IL: University of Illinois Press, 1964), 148.

13. Roger Launius, "Perceptions of Apollo," *Space Policy* 21 (2005): 130.

14. James L. Kaufmann, *Selling Outer Space: Kennedy, the Media, and Funding for Project Apollo, 1961–1963* (Tuscaloosa, FL: The University of Alabama Press, 1994), 57.

15. Brynes, *Politics and Space*, 52.

16. Ibid., 78.

17. Herbert Gans, *Deciding What's News: A Study of CBS Evening News, NBC Nightly News, Newsweek and Time* (New York: Vintage Books, 1979), 48.

18. Buzz Aldrin and Wayne Warga, "Return to Earth," *Good Housekeeping*, October 1973, 212.

19. Brynes, *Politics and Space*, 55.

20. Robert Zimmerman, *Genesis: The Story of Apollo 8* (New York: Dell, 1999), 235.

21. Brad Perry's interview with author, September 21, 2004.

22. Ibid.

23. NASA, *NASA, Space: The New Frontier* (Washington, DC: GPO, 1963), 2.

24. Neil Maher, "Shooting the Moon," *Environmental History* 9 (July 2004): 528.

25. William H. Cohn, "History for the Masses: Television Portrays the Past," *Journal of Popular Culture* X, no. 2 (Fall 1976): 281.

26. Van Dyke, *Pride and Power*, 154.

27. Zimmerman, *Genesis*, 199.

28. Ibid., 236.

29. THC's Modern Marvels showed a clip of author Andrew Chaikin, who described the powerful Saturn V rocket by reminding audiences of Norman Mailer's trope: "Man had finally found a way to talk to God," after Mailer witnessed the liftoff of the Saturn V rocket at the Apollo 11 launch (7/21/04).

30. This phrase is borrowed from the book title, "Chariots for Apollo: A History of Manned Lunar Spacecraft," Courtney G. Brooks, James Grimwood, and Loyd S. Swenson, Jr. (NASA SP 4205, 1979).

31. Thomas Pynchon, *Gravity's Rainbow* (New York: The Viking Press, Inc., 1973), 590.

32. United States Congress, Senate, *Committee on Aeronautical and Space Sciences, Subcommittee on NASA Authorization,* May 2, 1960, 241.

33. Brynes, *Politics and Space*, 60.

34. Tambay Obensen, "'Chasing the Moon': Watch Civil Rights Protest at 1969 Moon Launch in Rare Clip—Exclusive," *IndieWire*, accessed June 20, 2024, https://www.indiewire.com/features/general/chasing-the-moon-pbs-ralph-abernathy-1202153386/.

35. Ian Crawford, "To Still Boldly Go," *Prospect*, March 2003, 19.

Chapter 7

Pragmatism

Public excitement for the Moon missions began to wane after Apollo 11, coinciding with diminished political support and a reduced budget for NASA. Lessened public support for Apollo yielded fewer television audiences. The result was that NASA's rhetoric morphed into pragmatic discourse to justify the mission's mounting costs. *CBS Evening News* followed the NASA formula of emphasizing the utility of space science in hopes of piquing audience interest with "gee whiz" illustrations of lunar discoveries. In its reports of the practicality of space exploration, CBS mentioned scientific utility even as early as Apollo 11 when Eric Sevareid said the goal of that mission was an ancient one, "to unravel the riddle of the universe" (CBS, 7/21/69). With increased practical value came discussions of the scientific and technological benefits derived from such exploration. Each mission thereafter promised to be more scientific, more challenging, and more exciting than the one before. For instance, calling Apollo 12 "the most scientifically valuable manned flight to the Moon" (CBS, 12/12/69), Cronkite summarized its accomplishments:

> [The astronauts] *planted* a *complex* package of *scientific experiments* on the moon, *collected* some hundred pounds of Moon rocks and *visited* the unmanned Surveyor space craft and *brought* back parts of it—all of this done with rare good humor and enthusiasm, . . . *pursuing the goal* of *full scientific exploration* of the Moon's surface. (CBS, 11/24/69) (emphasis mine)

But Cronkite was also careful not to bore the audience with too much science. Instead, stories of "science" and "adventure" were often intertwined to regale viewers.

As audiences continued to plummet, CBS tried to find ways to make perfunctory missions interesting and important. Despite its "normalcy," space exploration was still complicated and risky, something that would become evident with Apollo 13. Following this trend, CBS featured segments in which Apollo astronauts conducted NASA-planned scientific experiments because, as reporter David Schoumacher suggested, "NASA would like to convince the general public of the practical value of space flight" (CBS, 2/7/71). But as Apollo wound down with missions 16 and 17, Cronkite seemed desperate to retain his space audience. In his attempt to convince viewers that public support was still alive for the remaining two space shots, Cronkite emphasized novel aspects, often featuring footage of astronauts with similar persuasive rhetoric. Stated Apollo 16's Charlie Duke,

> A lot of people say it's just another bag of rocks. Well, it's not just another bag of rocks. It tells the total history of the evolution of our solar system, and to me, that's a stupefying thought—that you can go up and pick up rocks in their original form, many of them three to four billion years old. Now that's a long time. (CBS, 4/14/72)

If we could understand these dynamic processes of the Moon and the Earth, Duke reasoned, then we can better manage our own natural resources. Such persuasive techniques endured until the last mission. Sometimes, the language was philosophical in nature. On the day the Apollo 17 astronauts returned to Earth, Cernan explained that the United States had grown scientifically: "A fundamental law of nature—you must grow, or you must die. Whether that be an idea, whether that be a man, whether that be a flower, or a country, I thank God that our country has chosen to grow" (CBS, 12/19/72).

In addition to the crews' pleas to continue the costly ventures, CBS renewed its focus on the astronaut himself in its space coverage. NASA's strategic decision to add Dr. Harrison Schmitt to Apollo 17's crew, the first professional scientist in space who was not by trade a pilot, helped. As Morton Dean reported, Schmitt had "big time academic credentials. He puts an end to those barking that NASA is not about science and that NASA has given science a back seat to engineering and PR stunting" (CBS, 12/5/72). CBS singled Schmitt out as "the first scientist in space, a geologist, a bachelor astronaut" (CBS, 12/5/72). Pointing to the fact that Schmitt was single might have undercut the family man icon of earlier missions. Instead, the astronaut, freed of his heroic responsibilities, became a symbol of science.

Science indeed. Beyond the merits of Apollo 17 lay the achievements of the entire Apollo program: revelations about the solar system's history and man's future in that system (CBS, 12/18/72). Calling Apollo a "scientific detective," CBS reporter Morton Dean claimed it was "just the beginning" as he summarized the utility of the program. Some 600 pounds of lunar material had been brought back to Earth for 800 scientists to probe the Moon's "deepest secrets" (CBS, 12/4/72). He explained how it was formed and speculated that its formation could shed light on the Earth's genesis. The answers learned justified the program: lunar samples could sustain life; meteors had formed most of the Moon's craters; and Moon minerals differed from Earth's. Placing new facts together in "a giant jigsaw puzzle," Dean said, was a continuous process (CBS, 12/4/72). Only ten percent of Moon samples had been analyzed, while the rest were being kept in reserve awaiting new scientific procedures for future investigation (CBS, 12/4/72). This constant reporting on the utility of Apollo might have appeased those who fulminated against NASA for over-spending, and against CBS for over-reporting a program that had already run its course. As Norman Mailer lamented, "The fact that you have this program which is glamourous . . . , high adventure, and that this program resolutely determines to become [solely] technological and dull is part of the tragedy."[1]

Similarly, in response to scientists who indicated they had just as many questions now—at the end of the program—as they did at the beginning, Cronkite mollified viewers who shared those same sentiments:

> Historically, great leaps in man's knowledge have taken place not in one but in two steps. It was the five-year voyage of the exploring ship *Beagle* and the mountains of new biological data it produced which led Darwin 20 years later to his theory of evolution. Only when the new study of radiation began poking holes in Newton's physics did Einstein produce his theory of relativity. Science moves, it seems, when a mass of new data throws old theories into question, and a man of genius finally pulls together a new, deeper understanding. It may just be that Project Apollo is now a mass of new data waiting on a genius. (CBS, 12/4/72)

Pushing the scientific merits of the program, Cronkite's challenge reinforced the potential for future exploration. But CBS was also quick to point out that, while Apollo 17 signified the end of the program, it also represented a "beginning": the start of a new manned space program. Wernher von Braun, the genius behind the Saturn V technology, declared that "all good things have an ending" but insisted that the Shuttle program would help us learn more about space (CBS, 12/6/72). When Cronkite asked von Braun why the public had been so excited about the first steps on the Moon (Apollo 11), but only two years later became so blasé about Apollo 17, von Braun suggested the metaphor of marriage. "Well, I think it's the excitement of the new. I mean, it's like getting married, you know, being married. The love is still there, and the excitement is still there, but it's no longer the honeymoon" (CBS, 12/6/72). A second "honeymoon" to rekindle that "love" would come with the reusable Space Shuttle. CBS was clearly cultivating an audience for NASA's next grand venture.

In fact, CBS's rhetoric continued to follow a pragmatic or scientific narrative pattern in the 1980s Shuttle years when it recalled Apollo, perhaps due to the tragedy of the Challenger Shuttle accident that killed all seven astronauts aboard. Here, the *modus operandi* might have been to reinforce the notion that space exploration still provided science that would benefit humankind despite its risks. Interestingly, the Apollo 1 fire that killed Grissom, Chaffee, and White on January 27, 1967, did not resurface in the news post-Apollo until *after* the Challenger accident, which ironically occurred on January 28, 1986—one day and 19 years after the Apollo 1 tragedy. Nonetheless, CBS has continued to remember Apollo's pragmatic efforts with discussions of other contemporary space projects. For instance, before reminiscing about Apollo 11 on its 30th anniversary, Dan Rather spoke about the status of U.S. space exploration with mentions of the International Space Station, plans for another Moon mission, and a trip to Mars. In speaking of the feasibility of each, CBS reporter Bill Harwood said that future Apollo-like missions to the Moon would provide a "life support system and habitat for the base of Mars missions" (CBS, 7/16/99). Similarly, CNN also spoke of the practicality of Apollo to better explain NASA's current projects. Although there were 842 pounds of lunar samples returned to Earth, more than 80 percent of the Moon had not been studied, leading reporter John Zarrella to conclude that "much of the Moon still remains a mystery," a mystery that the Lunar Prospector would help solve (CNN, 1/5/98). Thus, on the one hand, it was as if a modern-day Apollo-type program had to be presented with new utility because the audience was already well-informed of trips to the Moon. On the other hand, this might have been the networks' way of repackaging the program to make it more interesting to its audience. Either way, the media depended on the familiar to retell Apollo's story.

PBS, on the other hand, was divided on the extent to which Apollo was pragmatic. *To the Moon*, perhaps because it was a two-hour documentary, spent far more time describing Apollo's pragmatic merits, including the 3.7-billion-year-old lunar rock collected that gave clues to the Earth's geological history. Calling Apollo 15 "the first extensive scientific voyage . . . that was a change from the space cowboys of earlier times" (PBS, 7/13/99), the narrative reflected the shift in priorities among missions. Scientific discourse dominated this segment of *To the Moon* with discussions of the Hadley-Apennine Mountain range and Imbrium Crater sites, and how these two geological formations contributed a "radical" new theory of Earth's creation—a far cry from reading from the *Bible* during Apollo 8. In fact, the only mention of Genesis in this part of the narrative was in reference to a rock—the oldest lunar sample to date—that the press dubbed "Genesis rock" and what *To the Moon* called "a jewel in the crown of lunar exploration" (PBS, 7/13/99). Whereas *One Small Step* (PBS, 1/25/78) indicated that the science obtained from the Apollo missions was disappointing, *To the Moon* provided numerous examples of how the lunar rocks that were returned by astronauts settled scientific debates pertaining to the Earth's origins. Just as there were disputes among scientists, so too would there be debates among the public via the networks: was the cost of Apollo justified in terms of the science yielded? While critics of the program maintained that it was not, one of the ways in which the networks persuaded the audiences otherwise was to boost the promising future of space exploration.

There is an additional trend at play here. After the 1986 Challenger accident, into the 1990s, and still today, discussions of Apollo's pragmatic efforts have shifted from scientific utility to emphasizing the environmental, geopolitical, and economic practicalities of human spaceflight. President Bill Clinton, for instance, reminded CNN viewers at Apollo 11's 25th anniversary celebration of the importance of being environmentally conscious:

> Because we are so very finite, our responsibility to our planet must not be limited. That's why NASA's mission to planet Earth is also a very important part of our future in space. We have to continue to monitor the global environment from space and to act on what we learn. (CNN, 7/20/94)

But prior to that, in addition to space "out there," in 1972 NASA added nature to its focus here on Earth, when it launched the first Landsat satellite that would help explain environmental problems such as soil erosion, air pollution, and urban sprawl. The space agency's "Mission to Planet Earth," which began in 1989 and now is called "Earth Science Enterprise," continues to track Earth's living systems and ecological changes, but on a global scale, using several space satellites.[2]

Geopolitically, although NASA has claimed that space exploration is a "potent symbol of American democracy,"[3] it has also encouraged international cooperation, not only for financial support but to build relations with other countries in hopes of achieving international solidarity. That said, CNN recognized that space travel had become more pragmatic in the geopolitical arena. "Our space explorations today are important models for cooperation in the new post-Cold War . . . the International Space Station, this permanent orbiting space laboratory to be built with the help of 14 nations" (CNN, 7/20/94). Most of the discussions of geopolitical uses of space exploration, therefore, are associated with the International Space Station program. However, as evident here, anniversaries of Apollo are opportune times to recount space expeditions in general.

Economically, space has been portrayed as a means to accelerate the development of technology to help advance the American economy. Clinton coupled discussions of space exploration with "our way of life" to illustrate the importance of such advancement:

> [The astronauts] have shown us we can harness the technology of space in areas from the economy to the environment to education. The information and technology, the products and knowledge that grew out of our space missions, have changed our way of life forever and for the better. (CNN, 7/20/94)

Similarly, while PBS's *One Small Step* very briefly mentioned that Russians and Americans have discovered the economic opportunities of space, it hinted that NASA would need to become more pragmatic in its space activities if it wanted space exploration to become economically and scientifically fruitful (1/25/78). This was particularly fitting given the shift in space rhetoric that would become more pragmatic in the 1980s Reagan years to promote the Shuttle program to the American public.

Finally, NASA has articulated several observations regarding the economic benefits of space travel, all of which the networks have reiterated to audiences. The most obvious, of course, is the support it offers to aerospace industries, as it especially did during the Apollo years, which are vital to the American economy, still today employing over two million people in the United States alone.[4] For example, as the Apollo program ended, CBS reported that half the jobs at Kennedy Space Center and hundreds of thousands of jobs at NASA contractor facilities such as Grumman and North American Rockwell would be lost (CBS, 7/29/69, 7/21/72). In addition, NASA has noted that global economic competition has "replaced the Cold War political and military competition between East and West."[5] Working with the private sector, the United States therefore requires the enhancement of technologies that NASA deems "critical" to national security: power, computing, robotics, nanotechnology, biotechnology, communications, networking, and materials.[6] Expanding this idea, THC featured Gene Kranz, who stated that our future depends on space exploration. "Our survival as a species and economically depends on it because the only way we're going to compete in this entire world is though technology" (THC, 7/21/04). While THC did not explain the science obtained in this particular broadcast, it was still evident that THC condoned future space exploration so that the United States would remain a technological leader in global competition.

CBS Evening News has more recently reported this notion of competing economically in space exploration. In its celebrations of Apollo's 50th anniversary, CBS took on a new angle: sustainability. Anchor Elaine Quijano spoke of returning to the Moon in 2024 with the Orion program. "Apollo was about landing on the Moon;" this time "it's about a sustainable, long term human space exploration program," explained Northrop Grumman's Frank DeMauro (CBS, 7/21/19). CBS then showed a mock-up of the Gateway habitat where four astronauts could work and live for up to two months. The network also portrayed businessman Jeff Bezos as one who would work in cooperation with companies and governments. "Apollo was such an audacious project that when it was finished, the resources to sustain it just weren't available," stated Bezos (CBS, 7/16/19). Using 1969 Apollo 11 footage, CBS's Mark Strassmann reported that NASA would return to the Moon, then onto Mars. He then featured a clip with Apollo 11's Mike Collins who said, "Mars makes Apollo look like child's play" as a computer animation showed humans working on Mars. It then cut to NASA Administrator Jim Bridenstine who promised, "The next time we go [to the Moon], we're going to stay," to harvest its resources and be sustainable (CBS, 7/16/19). Once again, CBS reported the NASA narrative that technology is worth the investment, and it used the successes of Apollo to propagate that promise. With that in mind, we turn now to how the networks have portrayed lunar exploration technology vis-à-vis Apollo.

NOTES

1. NBC Commentary, "Why is the Moon So Boring," *The Observer,* May 6, 1972.
2. Neil Mayer, "Shooting the Moon," *Environmental History* 9 (July 2004): 529.
3. NASA, *The Vision for Space Exploration*, NASA Printing Office, Washington, DC, February 2004, 21.
4. "Industry Impact, 2023 Facts & Figures: U.S. Aerospace and Defense," Aerospace Industries Association, accessed May 15, 2024, https://www.aia-aerospace.org/news/aia-releases-2023-facts-figures-data-highlighting-ad-industrys-return-to-pre-pandemic-levels/.
5. NASA, "Race to Space," accessed June 20, 2024, https://www.nasa.gov/stem-content/race-to-space/.
6. NASA, *The Vision for Space Exploration,* 21.

Chapter 8

Technology

A fourth and final narrative theme involves Apollo's use of technology, falling into four sub-themes: militarism, technological successes, the human aspect of Apollo's technology, and the entertainment value of technology. Each of these themes corresponds with NASA's purposes for the missions. For example, because the goal of Apollo 8, which Astronaut Frank Borman called an "audacious journey" (PBS 12/26/18), was to sally forth against the Soviets in the Space Race, naturally, military metaphors enhanced the perception that it had done so. For Apollo 11, the first Moon landing, the coverage focused on America's technological prowess. The central message here, of course, was "we won with our superior technology." Given the near loss of Apollo 13, the human component of technology became the story, both in terms of lives nearly lost and of the women and men who saved the day with their technological ingenuity. Lastly, the entertainment value of technology became paramount in the final missions as it rescued audiences from mundane coverage of pragmatic findings.

MILITARISM

The militarism theme employed warlike or military metaphors to express not only the Cold War origins of Apollo but also to appeal to Americans' sense of winning. As a government agency, NASA characterized the Saturn V rocket as an all-powerful, military-like "weapon" in the Cold War, and, obviously, CBS followed suit. It was no surprise, then, that as the first mission to the Moon, Apollo 8 coverage employed the most military metaphors to describe the program, especially since the focus in early missions was on beating the Soviets. To illustrate, prior to launch, Cronkite referred to the Saturn V rocket as an instrument "treated gingerly with the utmost respect because its 363 feet of danger," and as "a potential bomb with the force of a tactical nuclear weapon" whose malfunction would cause a "raging inferno with 2.5 million pounds of TNT" (CBS, 12/17/68). While the Saturn V rocket was "heavy as a small warship," even the Apollo module itself was a "blunt-nosed missile," said Sevareid (CBS, 12/23/68). Displayed on television amid reports of the failed American effort in Vietnam, U.S. technology could still seem superior. Converting weapons to the peaceful purpose of flying to the Moon perhaps persuaded the public that spending massive funds on rockets had not been in vain. After Apollo 8, however, military metaphors waned significantly, with only a few mentions throughout the remaining course of the Apollo program. For example, Sevareid described the "atomic mushroom clouds" of Apollo 11's launch (CBS, 7/16/69), while Buzz Aldrin compared lunar dust to "gunpowder."[1]

In post-Apollo years, the comparison of the Saturn V rocket to a nuclear bomb continued to be a prevalent metaphor, especially on THC, which is to be expected given the network's fondness for war programming. The references to war were partly to dramatize, but also to provide a historical explanation for Apollo's origin. PBS also indicated that Apollo was the result of Cold War rivalries between two superpowers that needed to demonstrate technological superiority. The Saturn V rocket became a "weapon," a "memorial not just to American technology, but to the Cold War." The "Moon could be *conquered*," PBS's narrator insisted; "the Moon, it seemed, would *belong* to America"; "*victory* was confirmed with Apollo 11" (PBS, 1/25/78) (emphasis mine). But there was a difference in how the war metaphors were presented. During the Apollo years, NASA's technology was described as powerful just by virtue of being American. Nostalgic programs did so as well but were more likely to focus on the sheer power of the technology, and military metaphors enabled viewers to revisit the anxiety provoked by the Cold War. *Triumph and Tragedy: Apollo 11 and Challenger* trotted out the predictable: Wallace opened the program by telling viewers that the United States "mobilized as if it were going to war" once the Soviets had launched the first artificial satellite into space (THC, 3/10/99). Other scenes were clearly warlike. Asked to describe the Saturn V that would provide an answer to Sputnik, on *Modern Marvels: Apollo 11*, author Andrew Chaikin called it a "monster" with "the thrust of over 140 jet fighters" while a clip ran in the background of burning buildings (THC, 7/21/04).

Even NASA engineers, in retrospect, seem to have enlisted in a war. *Failure is Not an Option* narrator explained how Apollo Flight Director Gene Kranz's childhood dream of flight was "not borne of laughter, but one of fire," spurred originally by the Soviets' detonation of a nuclear bomb in 1949 (THC, 8/24/03). "Overnight, Russia ha[d] made space the new battlefield in the Cold War," said the narrator. However, NASA would be "led not by generals, but by engineers" (THC, 8/24/03). To recapture the period, THC showed frightening images of rockets launching amid bombs exploding, soldiers marching, military aircraft flying, missiles firing, and a close-up of the Soviet flag. The narrator recalled,

> Russia already has nuclear weapons. Now they have the world's first satellite, an unstoppable machine flying over the United States. Can it spy or take pictures? Can it drop a bomb? Four weeks later, another Sputnik. A hostile power is advancing into space, perhaps to control the planet. (THC, 8/24/03)

To further dramatize, THC featured vintage footage of General Jimmy Doolittle of the U.S. Air Force, who predicted, "It is quite possible that an aggressor nation that dominates space will then dominate the world. We just can't let that happen" (THC, 8/24/03). At this point, THC cut back to the engineer, once an Air Force pilot, but now the new "general" in executing military technology. Here, the young Kranz was pictured standing next to his airplane while fighting in Korea. But the program was not about the Korean War; rather it pertained to serving the United States. Apollo astronauts and Mission Control engineers stood stalwartly against the evil empire. As Kranz said, "[Apollo] moved from being a challenge literally to a *crusade*. This was now our mission to win this *battle*" (THC, 8/24/03) (emphasis mine). Kranz would repeat this rhetoric in subsequent years on *CBS Evening News*, "We won this *battle* for space in this room, and we *captured* the high ground, and we did not *surrender* during our *tour*" (CBS, 8/5/17 and 6/29/19) (emphasis mine). Conversely, PBS credited the heroic astronauts, not NASA's engineers, for winning the technological war. *One Small Step* spiced up its narrative by casting the astronauts as "Cold War warriors, ready to battle communism in space" (PBS, 7/13/99). Similarly, during Apollo 8's 50th anniversary, Astronaut Frank Borman on *Apollo's Daring Mission* said, "I never joined NASA to explore space. You know, basically I was a military person, and it was clear to me that we were in a serious confrontation with the Soviets. I wanted to be part of winning" (PBS, 12/26/18).

CNN chose a different narrative path in an Apollo 11's 45th anniversary story. While anchor Miguel Marquez called the mission "heroic" and its astronauts "legends," he reminded viewers that it was time to make history again with new "finish lines ahead" (CNN, 7/13/14). These finish lines had little to do with the Cold War and instead Buzz Aldrin, interviewed by Marquez, talked about helping veterans, who were committing suicide at a stunning rate of 22 per day as the network ran footage of the American flag being planted on the lunar surface. The narrative morphed from claiming lunar territory to instead helping those who served fighting in the earthly trenches.

Either way, even in this post-Cold War world, cultural institutions such as media outlets provide comfort to viewers by reminding them that the United States still has the technological capability to win wars—hot or cold. Communicators accomplish this through stories of technological achievements.

AMERICA'S TECHNOLOGICAL SUCCESS

As expected, CBS hyped NASA's successes, catering to America's love affair with the latest gadgetry. Dorothy Nelkin, discussed in chapter 1, has indicated that terms ending in "-est" typically are used to herald advancements in science and technology.[2] For instance, Cronkite told Apollo 10 audiences, "You no longer get the feeling that you're looking at the biggest, mightiest, heaviest rocket on the face of the earth. It's more like watching a big familiar ocean-liner routinely preparing for its regular run" (CBS, 5/16/69). While Cronkite downplayed the technology, he continued to use adjectives ending with "-est" to describe it, indicating that, although his audience might have become accustomed to the grandeur of the mighty Saturn V, the function it performed was essential to the success of the space program. For the remainder of the program, CBS coverage of each mission was replete with "firsts," "the most," and "-est" adjectives. Even in later years, CBS's Mark Straussmann, in a broadcast leading up to Apollo 11's 50-year anniversary coverage, reminded viewers that astronauts assembled the world's "largest" and "oldest" lunar rock collection, which taught us about the solar system (CBS, 7/14/19).

While technology placed that which was once off-limits within reach, Apollo 11, of course, was presented as the apex of technology. A week before its launch, with Kennedy Space Center launch director Rocco Petrone at his side, Cronkite took the audience to the launch pad and described "the bird's" complexity with its nine million parts of *precise* functioning (CBS, 7/10/69). Cronkite also took viewers to Bethpage, New York, where employees at a Grumman aircraft plant showed audiences how the lunar module was built. To impress viewers and to drive home the complexities of NASA technology, CBS captured on videotape the assembly of millions of wires in the LEM that were said to resemble "a psychedelic superhighway" composed of 40 miles of wire and 12,816 connections (CBS, 7/11/69). As he interviewed quality control workers, Cronkite commented on the LEM's "phenomenal record of performance" in the Apollo flights (CBS, 7/11/69). Ironically, failures of technology did not harm the program; even the botched Apollo 13 mission was deemed a success. Not only were the astronauts returned alive, but the House increased NASA's budget. As David Schoumacher reported, "The failure of Apollo 13 ha[d] done more for this space program than any of its past successes" (CBS, 4/24/70).

After the Apollo 13 incident, however, CBS was far more likely to report problems associated with the Apollo flights, including astronaut mishaps and machine malfunctions. Regardless, CBS still spent far more time praising NASA's technology instead of criticizing the problems the agency created. The network would have surely lost credibility among its viewers had it not been forthright in its coverage. True to form, CBS merely reported Apollo's problems without offering any critical analysis of the costs involved in correcting errors.

This formula of glossing over NASA's missteps would continue in post-Apollo years. For instance, *Crisis in Space: The Real Story of Apollo 13* narrator Mike Wallace blamed the mission's oxygen tank accident on bad luck *instead of* faulty engineering.

> Until the explosion, there had been little or no attention paid to the flight number, unlucky 13. Even so, the number 13 kept cropping up in a series of eerie coincidences. Liftoff came at 1:13 p.m. Houston time; that's 13:13 military time. (THC, 6/14/97)

The reference to the explosion occurring on April 13, 1970, transformed the progress into pseudo-science. On the one hand, telling the Apollo 13 story from the standpoint of superstition might have merely served the purpose of adding a new dimension to the well-known story, especially after the release of the *Apollo 13* movie in 1995 starring Tom Hanks as commander Jim Lovell. On the other hand, it seemed natural that THC would place the blame for the accident on misfortune, because as Wallace suggested, "The early problems aside, Lovell had every reason to be confident; after all, this was NASA, the agency that had pulled off a high-tech miracle of landing man on the Moon" (THC, 6/14/97). At this point, THC reminded viewers of what NASA did right by showing footage of Armstrong descending onto the lunar surface. Nostalgia thus took several forms.

Juxtaposing "triumph" with "tragedy" made the narrative seem even more dramatic as Wallace retold these familiar stories of human space exploration. For instance, "Houston, we've had a problem" has become synonymous with NASA's technological failure in collective memory. But also remembered was the outcome of the problem. As the *Modern Marvels: Apollo 13* narrator correctly recognized, "A triumph of ingenuity and guts, the Apollo 13 mission is often called NASA's most successful failure" (THC, 1/22/01). The networks, therefore, portrayed the duality of failure and success as good for America: we learn from our mistakes, and in the end, we overcome. This oxymoronic formula of successful failure even applied to the tragic Apollo 1 fire where all three astronauts perished. Although the Apollo 1 spacecraft never left the ground, "it may have contributed more than any other flight to the goal of reaching the Moon" (PBS, 7/13/99). Again, the tragedy made the triumph, and in the end, NASA's technology won.

Not surprisingly, current-day representations have remembered Apollo as "one of the greatest achievements of all time" (CNN, 7/20/09) and as reaching the "finish line" (CNN, 7/13/14). Narratives constructed the story as what Americans "can do if [they] want to do it bad enough," as Apollo 17 astronaut Gene Cernan put it (CNN, 1/9/97). Hard work breeds success. Of all the contemporary remembrances of Apollo, THC was most likely to focus on technological success. The network's strategy, however, was to detail advances made *as a result* of Apollo. For instance, *Tech Effect: Apollo 11* focused on the Saturn V, of course, but also on the lunar lander spacecraft, computers, the lunar camera, and customized spacesuits (THC, 6/20/04). Above all, the spacesuits—a "technological miracle"—enabled the Apollo astronauts to operate all of the other technologies, including the camera, computers, and spacecraft. Here, THC discussed at length how the spacesuits were designed to allow freedom of movement while it refreshingly ignored the United States and Soviet Union rivalry. Probably the most complex garments created in America, the spacesuits were fascinating because they acted as clothing and machines simultaneously to adhere to the physical conditions of space environments. NASA personified high technology,[3] but the spacesuits gave technology a personal touch because they were worn by the individual—not the institution that created them.[4] This had become increasingly important as NASA and the networks continued to push for manned space exploration, especially after two Shuttle explosions in which 14 astronauts perished and the public began to question the need for human risk in space.

Television programs reviewed additional "staggering" technological advances in other (non-space) fields. Not only would the seven-pound television camera "revolutionize newsgathering," but the integrated circuit technology used in Apollo's guidance computers would later become the desktop computers used today, as *Tech Effect: Apollo 11* explained (THC, 6/20/04). Other technological developments, such as CAT scans and MRIs, direct results of digital photographic enhancement created to map the lunar surface, have completely transformed the practice of medicine. Calling the lunar module computer "a marvel of engineering before its time," *Modern Marvels: Apollo 11* featured Mission Controller Steve Bales, who compared the guidance computer unfavorably to "the little hand-held calculator that you could buy at Walmart" to illustrate how commonplace Apollo's technologies have become in American culture (THC, 7/21/04). Finally, the narrator reminded audiences that the Apollo 11 mission "shatter[ed] limited thinking, bring[ing] with it the promise that all things are possible" (THC, 6/20/04). On the one hand, THC offered insights into the technological feats that were created *as a result* of Apollo. On the other hand, it played into the myth that, because the United States went to the Moon, it could accomplish anything technologically or otherwise. The belief, however false, is that technology can solve all societal problems. How many times, for instance, has the public asked, "If we can fly to the Moon, why can't we find a way to . . . [fill in the blank]?" The networks have capitalized on this myth. To dispel it would mean fewer audiences because it is a recurring narrative that binds together society.

As illustrated earlier, Apollo is often commemorated in tandem with news of more current space missions and their technological feats, or for planned excursions that will undoubtedly dazzle audiences. To illustrate the former, when a piece of foam insulation hit the Space Shuttle Discovery, CNN's Paula Zahn interviewed Apollo 7's Walter Cunningham, who said that he was not in the least concerned because the reporting had been "too alarmist" (CNN, 7/28/05). Zahn also consulted on air with John Pike, globalsecurity.org director, who suggested NASA ground the Shuttle and instead build an Apollo-like replacement that could operate at less expense and would be less hazardous to its crew (CNN, 7/28/05). A few years later, news of the successful docking of the Space Shuttle Endeavor at the International Space Station was tied to the success of Apollo on its 40th anniversary (CBS, 7/17/09). When SpaceX launched another Falcon 9, CBS ran footage of an Apollo liftoff from the same Launchpad 39 at Kennedy Space Center (CBS, 3/30/17). Also tied to Apollo's success on its 50th anniversary was the story of Richard Branson's Virgin Galactic, which occurred on July 20, 2019. Said Branson, "without the inspiration of Apollo 11, Virgin Galactic would not exist" (CBS, 7/20/19). In addition, introducing NASA's newest space excursion Artemis, Norah O'Donnell opened her July 16, 2019 news segment with, "America's new hope to return to the lunar surface is called Artemis, the Greek goddess of the Moon and twin sister of Apollo," as she remembered Apollo's 50th anniversary while talking about the future of space travel (CBS, 7/16/19). Similarly, footage of Bezos's Blue Origin flight fell on the 52nd anniversary of Apollo, so naturally CBS provided historic footage from 1969 (CBS, 7/20/21), just as it did for the Orion spacecraft splashdown report, 50 years to the day after Apollo 17 landed on the Moon (CBS, 12/7/22). As predicted, CBS replayed the 1972 newscasts and reminded viewers that it was the last time humans were on the Moon. Every memory of Apollo was an occasion to applaud its technology.

HUMAN COMPONENT OF TECHNOLOGY

Steven Goldman wrote that technology has traditionally been recognized in popular films since the 1920s as "a social process liable to selfish manipulation by powerful organizations."[5] On the other hand, engineering (and therefore the engineers who provide the technology) has been portrayed as "a body of objective technical knowledge capable of improving human existence and perhaps even human nature."[6] To be sure, the calming effect of engineering has often softened public suspicion toward new technologies. This was certainly the case with NASA engineers. As the debates that question the utility of space exploration and the so-called improvements that such exploration creates continue, the networks have consistently portrayed engineers as liaisons between a dubious public and the insensitive government apparatus that provides the technology. Thus, the networks have emphasized the importance of the human component of Apollo technology by using three discursive patterns. The first recognized the necessary and important role humans – men and women – played in creating the technology that would fly man to the Moon and return him safely to Earth, presenting personal stories of NASA's engineers and contractors. The second focused on the American ideal of teamwork—the cohesion of individual members for the betterment of the community itself, while the third pattern pertained to tensions between man and machine that arose when technology failed.

NASA and Contractor Engineers: Technology Workers

CBS realized early that, without the creators of the Apollo technology, there would have been no Moon shots. Many of its segments focused on the importance of engineers and scientists, especially as the missions progressed. NASA and CBS made an obvious attempt to convince the public that its budgeted dollars would be poured back into American communities through the thousands of space contractors NASA employed. At the Grumman factory in Bethpage, New York, Cronkite described in an 8.5-minute segment (of a 30-minute news broadcast) the importance of quality control, or human participation in managing the minute details of Apollo's technology. He returned there months later to illustrate how the lunar excursion module was built (CBS, 12/26/69) while engineers provided step-by-step accounts of how it worked in the weightlessness of space. Contractors were also consulted at CBS studios where they explained technological aspects of the mission—especially for the lunar rover, which was new to NASA and to the public. Always willing to humanize Apollo, Cronkite, who drove the lunar rover himself, asked a factory worker to demonstrate how the vehicle would work on the Moon (CBS, 7/23/71). This sort of "show and tell" served to keep the public interested in the mission while also informing them that the Apollo program employed many people behind the scenes, perhaps in a community much like their own.

Although CBS eschewed discussions of NASA's closed white-collar subculture in high-ranking positions, it did make a conscious effort to illustrate that space technology affected members of all socio-economic classes. Cronkite interviewed workers at Grumman and North American Rockwell, being certain to represent minorities that had been traditionally overlooked by NASA hiring practices. Having featured white-collar engineers in previous broadcasts, now the "guardians of the gate" (CBS, 7/11/69) were given the limelight. Cronkite interviewed several support staff including a white female and a Black male, in addition to two white males. The "lady electrician" told him, "I am very proud to be associated with Moon flight; it will be something to tell my grandchildren," while the Black male stated that, even if he got laid off, he would still feel like he was part of the program because it had become a part of him (CBS, 7/25/69). Both white males said that the experience of working on the Apollo program was worth the risk of being laid off but did not fear unemployment because the Moon landing was only the beginning of space exploration. Thus, even though Rockwell would soon reduce its workforce from 34,000 to 20,000 (CBS, 7/25/69), CBS wanted the public to know that all would be well—even if jobs were lost, being part of history was worth the price, and that the prospering American technology would always have a need for workers, blue- or white-collar.

During the course of Apollo, the network had become even more dependent on NASA contractors—the elite workers—especially in its stories of the ill-fated Apollo 13 mission. Reporter Nelson Bennett invited the audience back to Grumman, where engineers explained NASA's contingency plan through its simulators (CBS, 4/14/70). Whereas astronauts and politicians provided CBS with useful information pertaining to Apollo 11, engineers became the experts of choice to explain the efforts underway to get Apollo 13 home safely. Clergy and politicians like Richard Nixon publicly sang the engineers' praises (CBS, 4/17/70) as contractors Grumman and North American Rockwell continued to be spokespersons for NASA throughout the remainder of the Apollo missions. After the Apollo 13 near tragedy, CBS routinely used these experts to allay audience anxieties. With the exception of former astronaut Wally Schirra, along for the last Apollo mission (CBS, 12/6/72), more scientists than engineers were interviewed, perhaps to profess Apollo 17's focus on science. However, as CBS wrapped up the final mission, Cronkite again returned audiences to Grumman Aerospace. This particular segment unveiled how—now that the program had ended—one-half of the jobs would be eliminated at Kennedy Space Center and still more at contractors such as Grumman. Cronkite took viewers to the homes of aerospace employees who woefully described how losing their jobs would impact their families, forcing a move to seek employment. The segment and newscast ended with the sun setting behind the launch pad, symbolic of the lights going out on the program (CBS, 12/21/72), and again reminding the audience of the human component of the space program.

Enlisting the human angle may also have enabled CBS to illustrate NASA's role in creating jobs. Reminding the public that the lucrative aerospace industry employed many people, CBS communicated the notion that space exploration was good for the economy and thus the American way of life. According to CBS, the promise of the "American Dream" was as prominent toward the end of Apollo as it was at the beginning. A typical American boy could achieve his dream of working for NASA because he worked hard to save his money and did well in school, as could immigrants, as discussed in previous chapters. Convincing the public of this became even more necessary in the late 1960s and early 1970s as the counterculture pointed out the myth of the American Dream—a dream *not* available to all Americans.[7] This was evident in the networks' avoidance of criticizing NASA's employment practices. It would also be the focus of the 2016 blockbuster film, *Hidden Figures*, based on the book by Margot Lee Shetterly.

During the Apollo years, CBS featured segments of women and African Americans working as *support* staff on the Apollo program. Women, for example, skillfully stitched astronauts' suits, and African Americans were portrayed as apt employees working on lines to assemble the lunar module and other spacecraft. This was the network's way of involving minorities just enough to support the program and portray them as part of the "space community." At that point in time, becoming an engineer might not have been commonplace for minorities, but they now had the right to vote, and NASA needed taxpayer support regardless of sex or race.

In *Right Stuff, Wrong Sex: America's First Women in Space Program*, Margaret Weitekamp describes how John F. Kennedy urged girls to become involved in space exploration in a 1960 *American Girl* magazine feature page. Entitled "Lots of Room in Space for Women," the article indicated that the skills of men *and* women were needed in engineering. However, these invitations were in direct opposition to NASA's policy that time and again halted women's efforts to train as astronauts during the Apollo planning years.[8] It was acceptable for women to be "human computers,"[9] but in no way would they become the faces that would represent NASA. A more critical analysis of the struggles of dozens of highly qualified women pilots who took presidential rhetoric seriously and sought to become involved in every aspect of the space program, including becoming astronauts, would have undoubtedly provided a fresher perspective for the viewer. Indeed, to know this part of America's history would help explain Cold War tensions. Lyndon B. Johnson feared that women in space would "detract from NASA's missions and undermine its political support."[10] Perhaps the networks thought that offering a critical insight into NASA's unsavory practices would upset a predominantly male following of space-related stories and threaten their relationship with the agency that provided the material needed to sustain a number one rating in delivering Apollo news.

The same holds true for African Americans. On the one hand, space exploration was portrayed as the means by which the United States would obtain and maintain democracy. The networks pitched the Apollo program by way of democratic metaphors such as frontierism and peace. On the other hand, such ideals were not available to all—certainly not minorities outside of employment as support staff. As Lynn Spigel has noted, the only place that space-related multiculturalism appeared in the 1960s was on television's *Star Trek*.[11] A segment featuring an ousted Col. Edward J. Dwight, the first African American astronaut-candidate, would have provided the viewer a vista into NASA, and particularly civil rights struggles. Like women, qualified blacks—such as Dwight, whom NASA dropped from its program—had dreams of becoming astronauts and engineers. One would have to dig deep into print archives to discover the vast number of black engineers who contributed to the Apollo program. Examples include Carl Echols, a Grumman engineer who devised a plan that assisted in the safe return of Apollo 13 astronauts. "Ozzie" Williams, an aeronautical engineer, was Grumman's *only* liquid fuel rocket expert. Dr. Leon Moss was instrumental in designing the structure of the lunar module, while chemist Edwin Foster developed ways to keep fuel lines clean in the Saturn rocket for North American Rockwell.[12] The list is long. Instead, the networks have focused on white American- or European-born male scientists. To have televised the struggles these astute African Americans must have endured—their own humble beginnings and challenges in a world of segregated schools and separate drinking fountains—would absolutely have offered a new narrative to the same old Apollo story. To do so, however, would cast NASA as a failure in its promise of providing one giant leap for [*all* of] mankind. The networks would be damned. While Eric Hoffer said Apollo signified "the triumph of the squares"[13] of middle-class America, those in the African American community viewed it as one small step for "The Man."[14]

In post-Apollo years, networks continued to highlight the importance of human involvement in space exploration by giving the engineer a voice. People, of course, relate better to humans than to machines. *Failure is Not an Option*, based on Flight Director Gene Kranz's memoir of the same name, "add[ed] an important new perspective to the story of the Space Race" by examining "the 'whiz kid' atmosphere" of Mission Control and the enormous responsibility of its young members during the Apollo years.[15] Setting the mood for the program, the opening lines were, "They put men into space, but never left the Earth. They flew every flight from right here on the ground. . . . They took the dream of space flight and made it reality" (THC, 8/24/03). These two passages immediately enabled the viewer to place the members of Mission Control in the role of the astronaut—especially by employing the metaphor of the engineers flying from the ground. Borrowing a metaphor from another hero who also flew to protect the world against evil, Kranz said, "You cannot operate in this room unless you are Superman, and whatever happens, you are capable of solving the problem" (THC, 8/24/03). Indeed, the metaphoric use in this production replaces the heroic astronaut with the adroit engineer.

However, the face of the engineer had drastically changed. *Failure is Not an Option*'s narrator explained that, although Kranz and others who "pioneered the art of flight control" have retired, "the torch has passed to a new generation of Mission Controllers, cut from the same cloth, and viewed with the same spirit" (THC, 8/24/03). At this point, the new faces of Mission Control were shown, but this time the picture included African Americans and women. To close the segment, the narrator brought the story of Mission Control into perspective by indicating that these men and women had their own challenges (as the camera zoomed in on Space Shuttle Columbia), just as the original men of Mission Control endured the disasters of Apollos 1 and 13. Indeed, one of the lasting criticisms of NASA has been the absence of minorities in its workforce. Because they lacked diversity, NASA employees were often ridiculed for being "Leave it to Beaver" types: middle-aged, white, nerdy "yes" men. This was a stereotype left over from the Apollo era that many of the engineers wished to dispel, and the networks seemed happy to oblige, although they did not in any way offer an analysis of NASA's discriminatory practices at that time in history.

Decades later saw not only more women in NASA's tech arena, but also in news reporting, especially in coverage of Apollo 11's 50th anniversary. On July 15, 2019, her first day on the job as anchor for *CBS Evening News*, Norah O'Donnell opened the program,

> 50 years ago tomorrow, three men in spacesuits set off on the greatest scientific adventure of all time. As the astronauts of Apollo 11 headed to the Moon, the women of NASA were blazing new trails on Earth. Poppy Northcutt, JoAnn Morgan, and Margaret Hamilton were trailblazers, hidden figures critical to the success of the Apollo 11 mission. (CBS, 7/15/19)

As the narrator continued, "they were accepted in a profession not used to women," footage of the three women when they were young professionals rolled. The network then indicated that JoAnn Morgan was the only woman inside the firing room for the Saturn V launch, as she said that her director wanted her there to "listen for problems" (CBS, 7/15/19). CBS showed a still photograph of a very large group of men with Morgan's image highlighted because otherwise a viewer would not have known she was there amid a roomful of men. CNN showed this same highlighted photograph as reporter Randi Kaye told Morgan's story. When Kaye asked her how she ended up the only woman in the firing room during the launch of Apollo 11, Morgan said that she was told by her director that she was the best communicator. She later discovered that he had to convince Kennedy Space Center Director, Dr. Kurt Debus, that having a woman in the room was going to be okay. Said Kaye, "Do you think he had to go all the way up to the top for the men that were in the firing room?" Morgan replied, "Oh, heck no. I don't think so," to which Kaye added, "Exactly," at which point both women laughed (CNN, 6/21/19).

The segment portrayed the sagacious women as troubleshooters, blazing new trails for women in technology fields. Margaret Hamilton said, "We had to solve things that had never been solved before," for it was her group that wrote the lunar lander's guidance software whereby they "had a plan for every contingency" (CNN, 6/21/19). She was referring to the Apollo 11 alarm problem that earlier broadcasts had dramatized for effect, calling it the mission that almost didn't happen. She added, "There was no need to abort. Her software was fine" as the footage then turned to the clip of Armstrong walking on the Moon with his "One small step for [a] man" mantra (CBS, 7/15/19). The clip implied she had a hand in that "small step."

But the stories did not end there, for not only were these women groundbreakers, but they were also inspirations to many around the world. On CNN, Kaye talked about Morgan's love for science and learning while growing up, interning at the U.S. Army Ballistic Missile Agency at age 17, and then going on to become the first female engineer at Cape Canaveral (CNN, 6/21/19). Just like in *Hidden Figures*, Morgan had to walk to a different building to use the restroom, according to CNN, as the network played the famous clip from the movie of Taraji Henson, who played Katherine Johnson's character, walking a long distance to the bathroom on a rushed break. As the CNN clip progressed, viewers learned that evidently some men felt threatened when Morgan first started working in the firing room as she received obscene calls. "It was not going to be an object. I had too much fearlessness in me to let that be any kind of deterrent," stated Morgan (CNN, 6/21/19).

Coworker Roy Stark sat next to Morgan in the firing room. "We're all men. And JoAnn was there, and you know, she was a looker. You could never pull anything over on her because she would take and cut you to pieces because technically, she was extremely confident." While some men did not want her there, he continued, "there was no doubt about it. She had the moxie of what it took to be in the position of the only woman in the firing room for Apollo 11" (CNN, 6/21/19). Yet the most telling line came at the end of the news clip when Morgan said that she and her family watched the Moon landing event on television, like most everyone else. As footage showed a family gathered around a television set watching the by now familiar astronauts Armstrong and

Aldrin, Armstrong was heard saying, "That's one small step for [a] man, one giant leap for mankind." Morgan said, "My husband looked at me and said, YOU are going to be in the history books!" The story ended with Kaye affirming that, after Apollo 11, Morgan continued to break barriers (CNN, 6/21/19).

CBS also addressed the influence these female engineers had on *all* of humankind. For instance, "Poppy Northcutt, the only woman inside Mission Control, calculated the maneuvers that would bring those astronauts home" (CBS, 7/15/19). Following footage of Apollo 11's splashdown and the three astronauts in dinghies to be brought ashore, Northcutt then added,

> It was also an opportunity, I felt, to encourage other women to go into science and technology and understand that women could do these jobs. I got letters from all over the world from little girls and boys saying, "I didn't know women could do this,"

to which O'Donnell ended the segment with, "And women can" (CBS, 7/15/19). A couple of months later, O'Donnell reported that NASA showed off two new red, white, and blue space suits for upcoming Moon missions that would be more flexible for maneuvering in the capsule. "And guess what? It will fit women better! That's right," O'Donnell exclaimed as the network featured footage of a woman wearing the suit (CBS, 10/15/19).

In comparable fashion, PBS's *Apollo's Daring Mission* (12/26/18) also featured Margaret Hamilton and Poppy Northcutt explaining the complex technical aspects of the computers and software required to get Apollo 8 astronauts to and from the Moon safely. According to this documentary, the work of these women contributed to this risky mission's success. Hamilton, for example, developed the software that would protect Apollo's computers from errors, such as when astronaut Jim Lovell entered the wrong key sequence on the return flight. Whereas previous television and documentary narratives focused on Apollo 8's Christmastime reading of the Book of Genesis and the discovery of the Earthrise and Earth View photographs, *Apollo's Daring Mission* focused on the importance of computer and software technologies. Driving home the women's contributions in computer engineering, according to the narrator,

> The legacy of this overlooked mission is profound. Of all the Apollo technologies, perhaps the one that touches more of us in our everyday lives than any other is its pioneering computer . . . With its DSKY and guidance computer, Apollo paved the way for keyboards, mice, touch screens, computer-controlled airliners, factories, smart phones and more. (PBS, 12/26/18)

Interestingly, the producer of the documentary, Sue Norton, and the associate producer, Lisa Goodfellow Bowe, are women.

Similarly, a few years prior, and following yet another script to tell the Apollo story, CNN offered a news segment about Megan Smith, who left a high-powered position at Google to work for the Obama administration in hopes of getting children, especially the disadvantaged, interested in math, science, and technology. She was “a woman on a mission for other women, past and present,” the narrator explained (CNN, 7/4/15). As NASA mathematician Katherine Johnson’s picture was shown, Smith explained to viewers that it was Johnson who calculated the trajectories for Alan Shepard, John Glenn, and the Apollo astronauts. Aged 90 at the time, Johnson lived in Maryland and was “in an Apollo movie [*Hidden Figures*]. We’ve lost that history, and we need to stand it back up.” The interviewer replied, “There are not enough women in tech, and there are not enough African Americans in tech, right? I mean, this is a boundary you want to help break down?” The camera then cuts to an African American teenage girl.

> Of course it’s a very, very important issue for me. I definitely have led my own workshops in making sure girls of color are getting the exposure at least in computer science, because if you can’t see it, then you can’t aspire to be it.

The camera turns back to Smith. “Yeah, it’s definitely an ‘I’ll take this place over Google!’” (CNN, 7/4/15).

Yet it was the news crawling across the screen on the ticker while CNN segued into Katherine Johnson’s story that drove home the message. The headline read, “Punishment for going to school.” The text continued, “Men in Afghanistan throw acid in faces of girls on their way to school,” while the ticker scrolled,

> Two men on motorbike threw acid at three teen girls, official tells CNN. The three girls, ages 16 to 18, were hospitalized; two were in critical condition after attack, hospital says. Attackers reportedly told the girls they were being punished for attending classes. Authorities are working to find the attackers, police spokesperson says. (CNN, 7/4/15)

This horrific news, juxtaposed with an inspirational story that encouraged girls to become involved in STEM initiatives, evoked the message that at least American women have the freedom to pursue such dreams, even though their stories were not told until decades after Apollo had ended. It also underscored the need for gender diversity in a traditionally male-dominated industry.

Apart from women’s space contributions in engineering and mathematics, contemporary representations might not have always critiqued NASA’s antiquated employment policies, but they did present a story of those who fought the system to rise from nothing to putting the first man on the Moon. Introducing the audience to likely unfamiliar people who made the lunar shots possible, *To the Moon* called these men "visionaries dreaming remarkable dreams" (PBS, 7/13/99). They reportedly were already planning trips to the Moon in the late 1950s at Langley Research Center in Virginia where Apollo began. By enlisting the experiences of two particular engineers, PBS not only created a dramatic story, but it also offered a new perspective into the human component of the makers of technology. Dr. Maxime Faget, for instance, became a "legend" in U.S. technical circles for he created the entire fleet of American manned spacecraft (PBS, 7/13/99). Dr. John Houbolt, on the other hand, was an engineer who was largely unknown to Americans although he "fought" for NASA to hear his plan that would get man to the Moon, meeting Kennedy's deadline (PBS, 7/13/99). *To the Moon* described in detail Houbolt's "Lunar Orbit Rendezvous" (LOR) that would enable NASA to land two astronauts on the Moon while a third orbited above. This would be far more economical than landing the main spacecraft there, in which case the fuel requirements would have made such travel prohibitive. PBS called his story a "struggle" as it tried to persuade the viewer that Houbolt repeatedly attempted to convince NASA to adopt his idea instead of the rival plan of the great rocket scientist Wernher von Braun's "Earth Orbit Rendezvous" that would use two smaller rockets (PBS, 7/13/99). Because von Braun was a member of NASA and Houbolt was not, this notion of an "outsider" to NASA's elite further dramatized the narrative. Houbolt's effort could also be interpreted as an individual fighting the establishment. Because individuality and the preservation of freedom are two of the most enduring news values, sociologist Herbert Gans has also recognized that news "continually deals with forces that may rob people of their initiative as individuals."[16] PBS's portrayal of Houbolt falls into this category.

Houbolt, according to *To the Moon*, begged NASA to accept his idea. Bypassing NASA's chain of command, Houbolt risked his career by approaching NASA Associate Administrator Dr. Robert Seamans. Upon reviewing Houbolt's plan, Seamans decided to try it, especially given Kennedy's fast-approaching deadline of 1970, and thought that Houbolt's plan might actually work, despite NASA's top engineers and scientists adamantly opposing it. Even though Houbolt's LOR was accepted, according to *To the Moon*, he never received proper credit. However, according to NASA documents, not only did Houbolt receive public recognition,[17] he also received a monetary award from the space agency.[18] Immediately after *To the Moon* offered this part of the story, it featured Houbolt describing his humble beginnings as he was pictured writing complex mathematical equations on a chalkboard:

> I grew up on the farm, working 16 hours a day, milking cows in the morning under 20 [degrees] below zero and everything. And to me, to know that I've

> been involved with one of our greatest achievements of mankind, I feel rather special about that. (PBS, 7/13/99)

Thus, in much the same way that CBS and CNN portrayed the astronauts, and later how the engineers on THC were lionized, here PBS took an average citizen and told of how he arose from a mundane life and reached his destiny by getting man to the Moon, thereby fulfilling the American Dream, both for himself and his country. So, like CBS, CNN, and THC, PBS not only recognized the human component of technology, but also found a way to relate it to the American Dream.

Teamwork

Metaphors used to describe the engineers who designed and managed lunar spacecraft mirrored those attached to the astronauts, particularly the American model of teamwork to accomplish a common good. This "special breed of anonymous artisans, thousands of them" (CBS, 7/14/69) found a way to create a "clean, beautiful, successful act of human cooperation" amid the social strife taking place in American cities (CBS, 7/21/69). Not only was Apollo teamwork viewed as the model of organizational expertise and success, but American engineers, with their "Herculean labors" (CBS, 5/12/69), were cast by the media as naturals for "great human creativity" (CBS, 7/21/69). For example, Sevareid noted that some, such as author Jules Verne, knew a hundred years previously that Americans would be first on the Moon because, he said, "the Yankees are engineers just as the Italians are musicians and the Germans metaphysicians by right of birth" (CBS, 7/14/69). Yet another segment focused on Mike Vucelic, an engineer from Yugoslavia who came to America to be a part of the team that built engines for NASA contractor North American Rockwell. Vucelic liked to tell young people, "If I can go from designing gliders for the Yugoslav Air Force to designing a J2 rocket engine [for the United States], imagine how far *you* can go" (CBS, 7/25/69). Here Vucelic encouraged younger generations to be challenged and illustrated that one need not be born in the United States to achieve the American Dream. CBS then provided proof that immigrants can come to the United States and achieve success in science and technology. The network also acknowledged that the lunar landing might have been a U.S. feat, but without the help of several non-Americans over thousands of years, Apollo would not have occurred. They included "the ancient Egyptian astronomers, Galileo, Copernicus, Kepler, Newton, the early Russian rocket theorists, the British with radar, and the modern Germans and Russians," Sevareid said (CBS, 7/14/69). He then suggested that the plaque left behind by the astronauts, "We came in peace *for* all mankind," instead read, "We were sent in peace *by* all mankind" (CBS, 7/14/69).

Post-Apollo stories emphasized how individual roles were integral parts of the whole that enabled the system to work. THC, for instance, explained the team-like structure of Mission Control and how engineers performed critical activities—some concerning mathematics, some dealing with physics, and others pertaining to systems and mechanics. There were separate jobs for different kinds of thinkers, all of them fitting into the "culture" of Mission Control—"like a foreign country" with its own language (THC, 8/24/03). Here, the men laughed about the acronyms so typical of NASA-speak. But, careful not to make the men seem too "nerdy"—a stereotype associated with NASA employees still today—THC made light of the engineers' reputation of being geeks by showing images of them at their computers dressed in white shirts and slim black neckties with buzz cuts and most wearing eyeglasses. To contrast, THC juxtaposed images of hippies marching around the Haight-Ashbury district of San Francisco, sporting long hair, strumming guitars, and making peace signs while psychedelic music played in the background. "As the sixties unfold, Mission Control will be many things; one thing it will never be is hip," the narrator said (THC, 8/24/03). The engineers joked about having pocket protectors. Some said they had them, some bragged they did not, while still another admitted to having worn the symbol of uncool but added at least his was made of leather (THC, 8/24/03). Evidently, THC was making the point that, yes, the engineers were smart, but as a team, they also had a sense of humor needed to endure the high-stress job of Mission Control.

For CBS's remembrance of NASA teamwork and technology, there continued to be a focus on the engineer, not the astronauts. In a newscast that commemorated Apollo 11's 40th anniversary, while Barack Obama hosted Armstrong, Aldrin, and Collins at a celebration, according to the anchor, "But of course hundreds of thousands of workers helped get them there, including one whose quick thinking helped make history." Reporter Randall Pinkston described how Mission Control's Stephen Bales "had 15 seconds to continue or abort" the Apollo 11 mission. According to Bales, "I was scared to death" as he explained that he had to figure out how to make it work, as that was what he was trained to do (CBS, 7/19/09). In addition to the three astronauts, Bales was given a Medal of Freedom by Richard Nixon as the representative of everyone at Mission Control. CBS showed film of Nixon calling it a "technological miracle" that helped open "a new frontier in space" as the network displayed the Earthrise photo in closing (CBS, 7/19/09). Thus, once again, it was the technology workers who saved the day, perhaps to put a different spin on a well-rehearsed story of the astronauts.

Man vs. Machine

The Apollo technology, a paradox of mass and lightness, was crafted to land man on the Moon in the safest, most economical manner possible. Still, reporters employed an abundance of metaphors to describe Apollo's strange-looking machinery. While the Saturn V was merely "a monster" (CBS, 12/20/68), the LEM, or "Spider," said Cronkite, "looks strange, like something to step on" (CBS, 2/26/69), akin to "science fiction" (CBS, 5/19/69, 7/21/69). Because of its "out-of-this-world shape," the complex "creature" could operate only in space (CBS, 7/16/69). These ordinary metaphors might have also served to assure viewers that NASA's technology was benign and under control. This would be important, especially in the 1960s, one decade after the 1950s atomic bomb scares.

On a more basic level, although Sevareid realized that the United States "since World War II . . . had been the chief headquarters for theoretical as well as applied science" (CBS, 7/21/69), CBS reported a different kind of tension between humans and science and technology. Members of the public believed that man-made machines would unfold the secrets of the universe, which in turn would threaten to reveal secrets of the Divine. This was not a new worry for the space agency. John Glenn, for example, received several letters from people calling him a sinner because "God made people for the earth—not the sky or the moon;" "let God's world alone."[19] A few years later and less than a week after Apollo 11 landed, a rain storm inundated the northeastern seaboard, and weather reporters received phone calls from concerned citizens who said the storm was "divine punishment" for man going to the Moon (CBS, 7/29/69). Thus, for some, CBS's humanizing Apollo technology did little to convince them that the heavens were fair game in exploration or that technology was indeed safe.

These narratives also addressed the tension between technical failure and human failure. Sevareid noted that "every gadget represents a thought and a hand. There's no such thing as a technical success or failure, only human. It is the three [Apollo 11] men who fly tomorrow who are mysterious, not their equipment" (CBS, 7/15/69). After the Apollo 13 explosion, he ventured farther. Despite the "mechanical, mathematical perfectionism about these fantastic flights" (CBS, 4/14/70), they fail because in the end, humans—who control the machines—are fallible. Sevareid made this point by comparing the failure of Apollo 13 with other failures in the space program:

> John Glenn's danger when the re-entry device came loose, the narrow escape of Armstrong and Scott when the vehicle went into a violent spin, and of course the fire on the pad that took three astronauts' lives three years ago. Technical failures *are* human failures. Computers do *only* what men tell them to do. (CBS, 4/14/70)

Yet, despite these near-fatal malfunctions, U.S. technologies—and therefore Americans—always overcome in the end in mediated versions of history. This was evident in Cronkite's recollection that "Neil Armstrong, the first man on the Moon, once faced death in space in his wildly tumbling Gemini 8 flight four years ago" (CBS, 4/16/70). In this example, the near failure ended in success—the biggest success of all—Armstrong walking on the Moon. Despite its fallibility, Apollo 13 astronaut Swigert said he *still* had faith in NASA's technology, indicating that he would fly again because "that vehicle will be fully able to do whatever mission it is called on to do" (CBS, 4/24/70). The fact that someone who almost lost his life retained faith in technology more than likely reinforced the public's confidence in NASA, future Apollo missions, and, most importantly, American technology.

It is not surprising that networks post-Apollo continue to laud NASA's technology. For Apollo 11's 25th anniversary, CBS enabled NASA engineers to recall the decision to land on the Moon despite computer warnings to abort the Apollo 11 mission (CBS, 7/17/94). Here, CBS introduced new material possibly to prevent audiences from becoming complacent with the orthodox Moon story. Another possible reason might have been that, by now, CBS was competing with cable news networks which operate non-stop, thereby devoting more time to "untold" stories. In addition, cable has afforded networks such as THC two-hour time spans with which to delve deeper into the human dimensions of their stories. For example, the human component of technology also provided fodder for *Modern Marvels: Apollo* 13, which spoke of Apollo in terms of "a combination of human courage and advanced technology" (THC, 1/22/01).

In addition to Apollo's mechanical technology, *Triumph and Tragedy: Apollo 11 and Challenger* (THC, 3/10/99) addressed the technology of the human body itself and how it would react in unfamiliar territories. While most of the Apollo representations described throughout this book have pertained to man-made machines, the THC narrative addressed the somewhat forgotten story of man himself—not as a hero, but as a complex machine. Again, this information might have been a way to add something new to a story that had been told hundreds of times before. Additionally, as Kaufmann has argued, media preference for celebrating man over machine might have persuaded the public to favor manned versus unmanned space exploration.[20]

ENTERTAINMENT VALUE OF TECHNOLOGY

While television news dates back to the 1950s, broadcasters recognized its entertainment value beginning in the 1960s.[21] This proved to be perfect timing. Because space travel involved risks—many unforeseeable—CBS often took a more jovial approach to Apollo technology. Certainly, one way in which NASA persuaded the public to support its programs was to emphasize their spectacular nature.[22] Described by CBS as "acrobats" who liked to play jokes on one another, the Apollo 7 astronauts themselves believed that NASA's insistence that they use cameras to give Americans the first views of humans in space was pure "show business" (CBS, 10/14/68). NASA understood the value of spectacle and realized early on that its missions would need to appeal to Americans' sense of entertainment and colluded with networks in foregrounding a show-biz type format, instead of a purely scientific one, to keep the public interested, if not amused.

Just as Apollo 10 was to return the first televised color pictures from space,[23] the flight itself seemed suited to show business. In various sequences during the flight, the astronauts illustrated how they ate and drank in space—in weightlessness—and on several occasions, Mission Control thanked them for the "show." The astronauts also inserted references to Peanuts characters Charlie Brown and Snoopy into the live broadcasts, perhaps to remind audiences that Apollo 10's mission was to "snoop" around the lunar surface, and "Charlie Brown" was Snoopy's companion and guardian.[24] The show business angle—"the carefree attitudes" of the astronauts and the "kidding" (CBS, 5/21/69) by Mission Control—undercut serious discussions of technology, something for which NASA would later be criticized. However, for the task at hand, it was necessary. The public did not understand such terms as "escape velocity," but it understood the concept of a dog and its master, and such allusions enabled the ordinary viewer to identify with the program without scientific terminology. What was interesting about the mascots, however, was the extent to which astronauts and media alike spoke of the machinery as though it possessed life-like qualities. For example, the crew referred to themselves as "the five of us on Apollo 10—Tom Stafford, John Young, Gene Cernan, Charlie Brown and Snoopy [the crew, and the command and lunar modules]" (CBS, 5/20/69). In another instance, as Cronkite described the technological difficulties of communicating when traveling around the dark side of the Moon, he reported that "astronauts Stafford and Cernan and Snoopy confirmed all was well" (CBS, 5/22/69).

Caricaturing the command and lunar modules, superficially a way of humanizing space exploration, also served to commercialize it. For even before man stepped foot onto the Moon, CBS Commentator Sevareid forecasted, "We can anticipate miniaturized replicas of that first pair of boots [on the lunar surface] selling in souvenir shops. A moonwalk will, as they say, humanize the whole space business and bring us back to ourselves" (CBS, 5/26/69). Rampant popularization was apparent in televised representations. The Apollo 10 astronauts radioed popular songs "Up, Up and Away" and "Fly Me to the Moon" while telling Mission Control that "it's almost like science fiction looking back at [Earth]" (CBS, 5/19/69). Such articles led TASS, the Soviet news agency, to characterize the Apollo missions as "nothing more than a new entertainment for American High Society" (CBS, 5/22/69). TASS lampooned "nobility such as Vice President Agnew and former Vice President Humphrey, who apparently have nothing better to do than to go to Kennedy Space Center to look for thrills during a launch" (CBS, 5/22/69). Such segments appeared amid reports of what Cronkite called "space spectaculars," predicated on "exciting" prospects of close-up shots of future landing sites (CBS, 5/23/69).

In addition, as a nation of sports enthusiasts, Americans undoubtedly appreciated NASA's and CBS's sports metaphors. For example, Cronkite and Schirra described how the launch of Apollo 14 had "the feel of a football Saturday" because of the nice, cool weather and "the football crowd attitude" at the Cape (CBS, 1/30/71). The spectacular aspects of Apollo made launches a kind of sport. Sports metaphors ritualized the mission—viewers came to cheer Team America (NASA) and its goal of winning the Space Race. Cronkite refuted the observation that Americans had become bored with the space program. How could this be when well over a half million people came to the Cape to witness the launch in person (CBS, 1/30/71)? Like any other sporting event, not only were ordinary Americans there as fans of the space program, but as in any televised spectacle, such as the Super Bowl or the Kentucky Derby, Hollywood's elite were present. Calling them "Hollywood and industrial royalty," Cronkite named a few of the VIPs—actors and politicians—just as he did for the Apollo 11 coverage, recognizing the "spectacle aspect" of the event (CBS, 1/30/71). As Apollo 11 astronaut Mike Collins recognized, "It's simply the American way, to view a televised spectacular and think of it as a Super Bowl."[25]

Predictably, CBS ballooned the coverage of Alan Shepard, who sneaked a golf club to the Moon where he hit golf balls as live pictures were beamed back to Earth for the Apollo 14 mission (CBS, 1/29/71). Again, CBS was careful to note that the astronauts mixed "fun with their various scientific assignments" (CBS, 2/6/71). Even so, Roger Mudd reported that "*Golf Magazine* made Shepard an honorary member of its All-American Team," and discussed what professional golfers such as Arnold Palmer, Sam McDowell, and Geno Cappelletti thought of Shepard's handicap and backswing (CBS, 2/6/71). Because sports like golf dominate American leisure activities, at least some of the public tuned in as if for the latest match rather than for scientific discoveries. Whatever the reason, for NASA and for CBS, it was mission accomplished, as long as there was an audience. However, as they would later learn, such spectacles angered those among the public who had not been convinced that the show was worth what it had cost to produce.

For viewers not interested in sports, coverage seemed to have appealed to a wide variety of other tastes. In another display of an extracurricular activity, astronaut Mitchell conducted a non-NASA sponsored experiment in Extra Sensory Perception (ESP). Although *CBS Evening News* showed pictures of the ESP cards (CBS, 2/9/71), Mitchell refused to discuss the results of his experiment (CBS, 3/1/71). This was likely NASA's decision since it had been highly criticized for the costly lunar golf outings. Similarly, two final "PR ploys" added to the spectacularity. The Apollo 15 astronauts took a "Sunday drive" (CBS, 8/1/71)—an American ritual—this time on the Moon in the lunar rover. NASA and CBS depended on the Moon buggy to boost ratings, which it did at first. In what was called another "PR stunt," the astronauts canceled the first stamp on the Moon, although NASA claimed it was not privy to the plan. Still, CBS portrayed the incident as a "brand new Apollo stamp, the inauguration perhaps of Moon mail" (CBS, 8/2/71), implying that it would not be long before sending mail to and from the Moon would be as ordinary as sending it via airmail across the world. These "lighter moments" became "a tradition" among the astronauts on television (CBS, 8/2/71). Perhaps CBS and NASA felt that such devices were necessary to keep the public interested instead of boring them with esoteric scientific rhetoric and mundane details that came with pragmatic reports. These actions might have also been reminders that America's way of life on Earth would soon be commonplace for life in space, even if they only temporarily revived public interest.

While more contemporary representations have also capitalized on the spectacle of space travel, networks today have instead focused on the role of television itself in space exploration and its technological abilities in providing entertainment for a worldwide audience. During Apollo 11's 35th anniversary, for example, CNN's Miles O'Brien featured a NASA celebration that honored the astronauts, and an "icon from another arena," Walter Cronkite, the Apollo doyen of network news who covered the U.S. space program from its inception (CNN, 7/20/04). The image of Cronkite holding up a model rocket is now considered quaint, an entertaining relic of bygone space coverage. By all accounts, he was the stabilizer in an uncertain time, during which "the parades of yesteryear when Tang ran through an entire culture's veins and Walter Cronkite spoke for everyone," as CBS's Jim Axelrod described (CBS, 9/23/05). When *CBS Evening News* reported Walter Cronkite's death, he was called the "Apollo 11 mission's biggest fan, and he wasn't afraid to let his enthusiasm show . . . From the earliest days of the space program, Walter brought the excitement, the drama and the achievements of spaceflight directly into our homes" (CBS, 7/18/09). Buzz Aldrin continued, "I think we all owe Walter a great debt" (CBS, 7/18/09), and added he would have liked to have seen Cronkite fly on a Space Shuttle mission to "reminisce about what spaceflight is really about in the golden words that only Walter Cronkite knew how to communicate to all of us, all of people around the world" (CBS, 7/18/09).

On the other hand, THC was the only network to acknowledge the television camera as an additional technological innovation instrumental in Apollo's success. *Tech Effect: Apollo 11* spoke of television's importance to a "spellbound audience" of a billion people who watched the lunar landing (THC, 6/20/04), a phenomenon made possible by a seven-pound camera built on a million-dollar NASA-Westinghouse contract (THC, 6/20/04). THC chose to remember this technology, not only because it added a fresh narrative, but also because it acted as a kind of metajournalism, a way by which television could tell stories about itself. It was television, after all, that cultivated worldwide interest in Apollo during a time when it was deemed a "vast wasteland," and it was television that could recount the marvelous Moon tale.

In the later years, revisiting television also became a way for *CBS Evening News* to take a different angle on Apollo's anniversaries. While much of the technology at the time was created to provide coverage of the Moon shots, by today's standards, it was of very poor quality. On Apollo 11's 40th anniversary, CBS's Katie Couric reminded viewers that half a billion people watched the mission on television, but the grainy images made it difficult to discern Armstrong and Aldrin.

> Now, 40 years later, a film restoration company hired by NASA has digitally enhanced the video. The images are much sharper and reflections can even be seen in Armstrong's visor. NASA had erased most of its original tapes from Apollo 11, so copies were used along with videotape provided by CBS News to produce a high-quality record. (CBS, 07/16/09)

Here, CBS becomes the repository of history, not NASA, in terms of televised images. In addition, even with the "new and improved" video, CBS replayed the old footage of Walter Cronkite's Apollo reports.

Apollo's 50th anniversary news provided a new perspective on how CBS reported the Moon shots. Standing beside an old television console from the era, Nora O'Donnell said, "The Moon has orbited the Earth 669 times since the Apollo 11 mission in July 1969," and most television sets were black and white, as only one-third of U.S. households had color television (CBS, 7/16/19). As predicted, O'Donnell said most people watched the historic broadcast on CBS with Walter Cronkite, as his face appeared on the screen of the old console. She then introduced footage of Joel Banow, whose work is described in chapter 4 of this book and described his groundbreaking work. "As the director, his job was explaining each stage of that very complicated eight-day mission with space-age graphic simulations that rivaled reality." Former CBS reporter David Schoumacher then poked fun at Moon hoaxers, "Today, there are people going around swearing that we never went to the moon. That it was all Joel Banow, did it all in the studio," (CBS, 7/16/19). Although humorous, his comment underscored the fact that it was the network's responsibility to illustrate a never-seen-before event as realistic as possible for its audiences.

In sum, televised representations of Apollo's technological feats from the 1960s to today cannot be complete without discussions of NASA and contractor engineers, astronauts, and technology workers, but also broadcast anchors, producers, directors, and artists. Since the beginning of the Apollo program, CBS News and later CNN, THC, and PBS recounted the lived experiences of an array of humans who had a tremendous impact on NASA's successes. In the end, this "technological puzzle," as Norah O'Donnell called it (CBS, 7/24/19), could not have happened without blue-collar and white-collar workers and those from all walks of life: genders, races, and cultures.

NOTES

1. Edwin E. Aldrin, Jr., "Lunar Dust Smelled Just Like Gunpowder," *Life*, August 22, 1969, 22.
2. Dorothy Nelkin, *Selling Science: How the Press Covers Science and Technology* (New York: W. H. Freeman and Company, 1987), 90.
3. Douglas N. Lantry, "Man in the Machine: Apollo-Era Space Suits as Artifacts of Technology and Culture," *Winterthur Portfolio* 30, no. 4 (1995): 220.
4. Ibid., 203.
5. Steven Goldman, "Images of Technology in Popular Films: Discussion and Filmography," *Science, Technology & Human Values* 14, no. 3 (Summer 1989): 282.
6. Ibid.
7. Peter Braunstein and Michael William Doyle (Eds.), *Imagine Nation: The American counterculture of the 1960s and 70s* (New York: Routledge, 2002), 3. See also Rebecca E. Klatch, *A Generation Divided: The New Left, the New Right, and the 1960s* (Berkeley, CA: The University of California Press, 1999), Todd Gitlin, *The Sixties: Years of Hope, Days of Rage* (New York: Bantam Books, 1987), and Paul Berman, *A Tale of Two Utopias: The Political Journey of the Generation of 1968* (New York: W. W. Norton & Company, 1996).
8. Margaret A. Weitekamp, *Right Stuff, Wrong Sex: America's First Women in Space Program* (Baltimore, MD: John Hopkins University Press, 2004), 134. For example, Johnson had written "Let's stop this now!" on a memo he was to sign to NASA Administrator James Webb, asking Webb if women had been discriminated against. According to Weitekamp, the memo had been buried for 40 years.
9. Ibid., 134.
10. Ibid., 90.
11. Lynn Spigel, "White Flight," in Lynn Spigel and Michael Curtin, Eds., *The Revolution Wasn't Televised: Sixties Television and Social Conflict* (New York: Routledge, 1997), 48.

12. Steven Morris, "How Blacks View Mankind's 'Giant Leap'," *Ebony*, September 1970, 34.

13. "The Moon and Middle America," *Time*, August 1, 1969, 11.

14. Lynn Spigel, "Outer Space and Inner Cities: African American Responses to NASA," in Lynn Spigel, Ed., *Welcome to the Dreamhouse: Popular Media and Postwar Suburbs* (Durham, NC: Duke University Press, 2001), 160–161.

15. "Failure is Not an Option," *The History Channel*, accessed April 29, 2005, http://www.historychannel.com, now https://www.history.com.

16. Herbert Gans, *Deciding What's News: A Study of CBS Evening News, NBC Nightly News, Newsweek and Time* (New York: Vintage Books, 1979), 50.

17. Letter from NASA Headquarters to John Houbolt. NASA History Office, Washington, DC.

18. Roger D. Launius, former Chief NASA Historian, conversation with author, Washington, DC, October 15, 2004.

19. John Glenn, *P.S. I Listened to Your Heart Beat: Letters to John Glenn* (Houston, TX: World Book Encyclopedia Science Service, Inc., 1964), 167–168.

20. James Kaufmann, *Selling Outer Space: Kennedy, the Media, and Funding for Project Apollo, 1961–1963* (Tuscaloosa, AL: The University of Alabama Press, 1994), 63.

21. Michael Schudson, *Watergate in American Memory: What we Remember, Forget, and Reconstruct about the Past* (New York: Basic Books, 1992), 133.

22. Mark E. Byrnes, *Politics and Space: Image Making by NASA* (Westport, CT: Praeger Publishers, 1994), 157.

23. Due to technical difficulties, this was not accomplished.

24. Richard L. Lattimer, *All We Did Was Fly to the Moon* (Gainesville, FL: Whispering Eagle Press, 1985), 61.

25. Michael Collins, *Carrying the Fire: An Astronaut's Journeys* (New York: Farrar, Straus and Giroux, 1974), 464.

Conclusion

Decades after the Apollo missions to the Moon, computer programmer Peter Shann Ford analyzed a tape of Neil Armstrong's famous line, "That's one small step for [a] man, one giant leap for mankind." For years, historians and grammarians have criticized Armstrong for leaving out the "a." Armstrong, however, had always contended that he pronounced the vowel; it was the unreliable 1969 transponder technology that had omitted it. Ford found evidence that the "a" was indeed spoken by Armstrong.[1] In response, some have asked, "We went to the Moon so who cares if the 'a' is there?" Good question, and one that gets at the very heart of this study: Americans want the memory of our nation's journey to the Moon to be as purified as possible. It would be sacrilege for something as trivial as a missing letter to ruin our collective memory of one of the greatest technological feats ever attempted by humankind, and the "greatest single broadcast in television history."[2]

This book has offered a rhetorical analysis of Apollo representations on CBS, CNN, THC, and PBS while examining the degree to which NASA's Public Affairs Office guided the rhetoric at the beginning of the program. My research began with the assumption that the Apollo program was presented to the American public through discourse that is more analogous to the ritual view of communication than the transmission view of simply communicating aspects of science and technology. Analyzed in terms of the ritual view, reporting on Apollo was more concerned with selling the program via the discourse of community and the sharing of symbols held dear by Americans at the height of the Cold War: faith in the system; love of God, country, and family; hard work and perseverance that bred success; and frontierism and exploration as especially American traits essential to surviving politically, socially, and economically.

First, it was necessary to examine the role NASA had in shaping network rhetoric. While NASA had a heavy influence on CBS, it did not dictate to the networks what to say. Rather, news personnel developed very close personal relationships with NASA employees who were bound by the Space Act of 1958 to be forthright in their dissemination of information. Even so, the space agency correctly recognized the entertainment value of the Moon shots. Similarly, broadcast journalists, bound by the traditions and conventions of newsgathering, also realized the mass audience their space stories would bring and worked with NASA to obtain the best information to deliver the exciting story. Thus, while no collusion was detected on either side, the cultivation of close relationships resulted in mutual benefit. Networks portrayed NASA and its programs in the best light while the space agency gave CBS a plethora of information because both were looking to attract a mass audience. This reciprocal relationship enabled CBS to lead in television audience ratings in Apollo's early years and to maintain that lead as the program progressed. There was also a bond of technological necessity: the networks were forced to rely on NASA for technical data, and NASA needed assurance that broadcasts were accurate. All of this exemplifies the Shoemaker and Reese model of media content, whereby individuals (such as Walter Cronkite) work within the boundaries of media routines (resources/technology) and around internal factors (CBS) and external factors (NASA) that act to portray the intended ideology (America is the technological leader).

In an analysis of NASA public relations, Mark Byrnes found that the agency's public statements during the Mercury, Apollo, and Shuttle eras followed three distinctive discursive patterns: nationalism, romanticism, and pragmatism. The research presented in this book confirmed these patterns and detected yet another—technology—while discovering that television networks mainly duplicated NASA's rhetoric but also adopted their own vernacular inspired by a fascination with NASA and its Apollo missions. Networks often borrowed NASA's metaphors verbatim, especially those involving national goals, American frontiers, and heroic deeds. By reiterating the space agency's rhetorical strategies, the networks catered to *hoi polloi*, for these were stories that enabled average Americans to relate to the intricacies of space science. Thus, although NASA had no direct control over the networks, it provided tangible promotional materials such as mission patches, press releases, education programs, and even astronauts, all saturated with NASA's ideology.

This author examined how the great enterprise was discussed on CBS News during the years that the Apollo missions took place, 1968–1972, and in post-Apollo years, 1973–2022, which marked the 50-year anniversary of NASA's last landing on the Moon with Apollo 17. I provided a detailed rhetorical analysis of how NASA and the networks translated these missions so that the public could understand space travel and furnished an account of the narratives each used to convince the public that the $25 billion adventure was worth the cost. Professional science narratives employ metaphors to describe intricate physical and chemical processes as they are even more necessary for a general audience struggling with complexities. For the networks, those metaphors lent themselves to a ritual communication mode.

In short, the Apollo story cannot be recounted without mentions of the American flag and John F. Kennedy, whose life symbolized triumph and tragedy, and who promised his citizens the Moon as the United States was in an urgent race during the Cold War. Networks portrayed "Old Glory" prominently as it symbolized inspiration, achieving the "impossible," unity, and an open and free society. These and other images of nationalism asserted the certainty of American success. The Space Race would be won, and Americans would share in the excitement.

Romantic metaphors burnished other rituals such as exploration, emotional rewards, challenge, inspiration, human spirit, spirituality, and satisfying curiosity. In addition, space was another American frontier, open to exploration by a cohort of heroes. Characteristic of the rhetoric used, television ritually rehearsed the virtues of the brave astronauts and self-made individualist engineers. Interestingly, the "hero" status dissipated when the rhetoric of Apollo shifted from "race" to "science," even though the missions were equally as dangerous after Apollo 11, as evident with Apollo 13. The romantic symbol of the hero had soon run its course, so scientific metaphors then served to convince the American public that the Moonshots had tangible value. This had become increasingly important with mounting complaints about the astronomical costs of future space travel. Therefore, after the initial Moon landing, the rituals expanded to include another American value: pragmatism. Here the metaphors underlined the social, political, and economic utility of space exploration—partly to justify expenditures, and partly to ensure confirmation and contiuation of the space program.

In addition to CBS's reliance on "gee whiz" scientific illustrations, as the flights moved toward completion, Apollo continued to be touted by CBS, employing adjectives such as "more," "most," and other superlatives to describe the missions. Interestingly, as technical and scientific complexity increased, public interest faded almost in direct proportion, and the use of metaphors dropped drastically. This was surprising because metaphors and other rhetorical strategies are often used to explain esoteric scientific terms to non-scientific publics.[3] In Apollo's case, it seemed as though CBS and NASA saw writ large that the public had had its fill, and thus decided not to press forward with a program that had already been exhaustively reported.[4] However, pragmatism provided a natural transition to Space Shuttle program coverage. The Shuttle had actually been formally proposed in 1966 (before the first Apollo mission flew), and publicly unveiled in 1968, the year of the first successful Apollo mission, Apollo 8. The vehicle itself was presented as reusable and therefore economical, factors that would convince the public that space exploration is affordable and sustainable. In any case, despite pragmatic narratives, there continues to be a dramatic flair to news reporting that serves NASA's objective of keeping public support for funding and CBS's goal of maintaining audiences. This trend continues with the reporting of current missions and news of tight Congressional budgets.

A fourth pattern—the invincibility of technology—was evident throughout the course of Apollo and into its aftermath. Not just a collective enterprise with national, romantic, and pragmatic aspects, technology emerged as a topic unto itself. American industrialism and post-industrialism emerged as elements of ritual celebration. Here, the networks focused on the success of America's technology: Apollo was equated with U.S. innovation. A variety of war metaphors communicated the power of U.S. technology, for it was believed that whichever superpower controlled technology would also evince superior ideology. It was obviously important, not only for ratings but also for future missions, that networks detail the magnificent technological advancements made as a result of the Apollo missions.

Despite touting America's technological prowess, the networks were diligent in illustrating the human component necessary for technological success, especially in their discussions of teamwork. Teamwork, a symbol of American organizational know-how, was illustrated via stories of successes *and* failures, especially in returning the Apollo 13 crew safely. NASA engineers' ability to work together for the common good saved the day. Faith in the system functioned to keep the American public hopeful of NASA's technological abilities and trustful of the U.S. government during a failed effort in Vietnam and in the streets of America. In the post-Apollo years, network coverage, particularly on CBS, while still touting NASA's innovations, has focused on futuristic technology develped by spacecraft manufacturers, framing the new "space race" as between entrepreneurs Jeff Bezos, Richard Branson, and Elon Musk to see who could reach the Moon, Mars, and beyond, since it is private enterprise that will fund future space travel due to limited Congressional budgets. Televised representations did this while recollecting the golden years of Apollo to remind viewers that at one time in history, reaching the Moon by the end of the decade was also a pipe dream.

Finally, in covering the technical aspects of Apollo, combining the medium with the "sport" of flights to the Moon ritualized the excursions as an American activity. Here, the networks publicly rooted for their team—NASA—to make safe journeys to and from the final frontier. Ritualizing the activity and providing common ground created a community-like sharing of symbols that garnered support for the costly project. Generally speaking, networks portrayed NASA ritualistically as a marvelous team and used sports metaphors to encourage spectatorship, fandom, and vicarious participation among the audience.

A final quest was to determine which Apollo narratives have withstood the test of time. Because "one of the most enduring news values is the presentation of the freedom of the individual against the encroachment of nation and society,"[5] networks, especially THC and PBS, were careful to tout the American ideal of the individual, the self-made man. What was different in remembering Apollo was to whom the rhetorical strategies applied: there was a shift in whom the networks deemed to be the Christopher Columbus *de jour*. First, it was Kennedy who proposed the lunar program, then the astronauts who acted as the frontiersmen conquering the vast regions of space. While the stories of astronauts' humble beginnings served to enable the average American to identify with them as "good ole boys," contemporary narratives are far more likely to feature engineers in the role of frontiersmen.

During the Apollo years, try as NASA and the networks might to present women and African Americans as significant participants, gender and racial inequities were visible everywhere as the narratives focused on the accomplishments of mostly white American or European-born males. This is not surprising since females were not admitted to NASA's Astronaut Class until 1978, along with the first African American and Asian American astronauts,[6] a decade after the first mission to reach the Moon's orbit with Apollo 8. Although THC and PBS were more likely to feature the accomplishments of white American or European-born men, current CBS, CNN, and at least one PBS remembrances of Apollo focus on the contributions of women who were instrumental in getting the astronauts to and from the Moon safely. Today's female news writers, producers, and anchors such as Norah O'Donnell might have been influenced by *Hidden Figures*, released in 2016, and added important narratives of women and people of color to pay tribute to Apollo's successes, and justifiably so.

For the male influence of space contributions, the cowboy narrative spanned generations, shifting from astronaut to engineer and then to producers and executors of technology who provide funding for the costly endeavors to new frontiers. For example, Mission Controller John Aaron, characterized as the American cowboy, was the archetypical American who opened the frontier, conquered a dangerous landscape, and settled the spacious West (or the Moon in this case). These narratives continued with images of billionaire Jeff Bezos being interviewed wearing a cowboy hat after his historic flight. The cowboy icon was meant to appeal to Americans' sense of respect and love of country, while also emphasizing the individuality of the engineer who had an appreciation of his humble beginnings with Aaron, and the monetary success of Bezos. In these instances, CBS, CNN, PBS, and THC used classic "rags to riches" stories to illustrate what ordinary men have endured to reach their goal in achieving the "American Dream." As Kaufmann has pointed out, "the media's ideal individuals successfully struggle against adversity, overcoming forces more powerful than themselves."[7] Today, the Space Race is now portrayed as being between cowboy billionaires, a contest of who will be the first to send civilians into space with tourism ventures and who can better assist the government with space exploration to eventually inhabit the Moon and Mars.

There are several possible reasons for the shift of metaphor from frontiersmen to technologists in the late 1990s and early 2000s. Engineers might have been portrayed as the new frontiersmen given America's technological push to compete in and dominate world markets. Some observers have claimed that the United States has been experiencing a "quiet crisis" whereby it is slowly losing its edge in scientific and technological innovation, slipping behind countries like China, South Korea, and India who are quickly taking the lead.[8] U.S. patent grants and publishing in science and engineering journals, two indicators of innovation, have steadily declined, leaving some to warn decades ago that "the breathtaking burst of discovery that has been driving our economy for the past half century will be over."[9] In fact, some experts believe that the United States has already lost its global leadership in science and technology.[10] Thus, providing audiences with the exciting, if not glamorous, side of technology might serve to pique an interest in science and technology by way of space exploration. And just as the frontiersmen narrative shifted from astronaut to engineer, so did that of hero—"unsung hero"—particularly in remembrances of the Apollo 13 story where engineers rescued the crew. Clearly, the networks were searching for new and different ways to tell the Apollo story as STEM initiatives abound in education sectors across the country, and the engineers provided that narrative.

An additional way in which the Apollo story was repackaged was in the very act of television telling a tale about itself. In *Covering the Body*, Barbie Zelizer argued that, in its reporting of the Kennedy assassination, the media promoted themselves as cultural authorities.[11] Applied to the task at hand, in his reporting of the space shots, Cronkite became a science authority to whom television audiences turned to explain the complexities of the race to the Moon, with the help of CBS Special Events Director, Joel Banow. Moreover, CBS used the Apollo story to boost its credibility and champion itself as a pioneer in broadcast technology. In the end, then, it became a cultural authority that created and maintained the symbolism needed to garner support for the Apollo missions. Years later, THC reinforced that American memory by coalescing vintage footage with current-day remembrances of the event. Thus, not only were networks such as CBS cultural authorities in years past, but today's networks credential themselves by keeping the tradition of television as storyteller and authority, past and present.

A final important and interesting finding for CBS's Apollo coverage from 1968 to 1972 was the extent to which CBS applied contradictory narratives to Apollo stories. Conflict is a necessary element in journalism.[12] In *60 Minutes and the News*, Richard Campbell offered an explanation as to why binary oppositions are necessary in news reporting. Stories that create problems—like binary opposites—fit into the mythic narratives Middle Americans hold dear. Storytellers, such as news organizations, attempt to resolve the problem at the end of the narrative, and it is at that point that social norms and values are reaffirmed.[13] Thus, binary opposites help distinguish the problem from the solution. Applied to the research at hand, there was an abundance of binary opposites. The astronaut was a "Cold War warrior," yet a frail mortal who came in peace for all mankind. Portrayed as a hero at first, the astronaut was later cast as an ordinary American. As a hero, he saved the United States from evil Communists by beating the Soviets to the Moon. When that narrative was exhausted, American social values were reaffirmed by the astronaut being portrayed as an average American with his faith and belief in God and country. While the mission itself became a binary opposite (race vs. science), the astronaut's role was once again transformed: as a scientist, he served his country by advancing it technologically. CBS's narratives and metaphors constantly shifted to reflect these binary opposites, creating meaning as the story progressed.

The contradictions applied to Apollo technology as well. Apollo would create a better way of life for humans, yet it came at a time when bombs and warfare destroyed life. Military metaphors, such as nuclear bombs, were used to describe the Saturn V rocket, but the rocket also enabled the United States to go in peace for all humans. Apollo 9 coverage pitted the drama of the mission against a show business aspect, which was in opposition to scientific discovery. Apollo 10 was described as historic yet routine. Of course, examples of binary opposites were bountiful for Apollo 11, the pinnacle of the program. America's trip to the Moon had scientific utility, yet it was a social substitute in many ways. It was portrayed as bringing peace between the United States and the Soviet Union, yet it was a substitute for not fixing the social problems on Earth, such as poverty, war, inflation, and social unrest. While Apollo's successes would place the United States as number one technologically, it was a way to repair U.S. foreign relations.

Similarly, the Moon was a romantic, myth-producing object, yet provided scientific evidence of how our Earth and solar system were created. Space exploration and Apollo were described as markers of a technological age, yet there was a focus on the human component of getting man to the Moon, and a spiritual revival because he went. Apollo 12 was pitched as normal, but historical; routine, but different; routine, but important; interesting, yet adventurous; exciting, yet scientific; fun, yet work-related. Binary opposites shifted again once the Apollo 13 accident occurred. At first, the program was described as a costly scientific venture that would produce more knowledge than the other missions. The rhetoric then shifted to a recovery mission to return the astronauts home safely. In the end, it was deemed a "successful failure," perhaps the most well known of binary oppositions. For Apollos 14, 15, 16, and 17, the binary oppositions became the ordinary versus the dramatic, entertaining yet scientific. This also might have been an attempt to dramatize the missions that had become commonplace, and to appeal to a wider audience (i.e., those who did not support the Apollo program) to affirm American social norms and values, such as longing to be number one in space exploration yet wanting to be monetarily conservative to fight other social ills. In each of the cases, presenting opposition served to enhance the public view of NASA while underscoring social mores. For instance, presenting Apollo 13 as a "successful failure" highlighted NASA's ability to overcome a life-threatening problem as it cast the engineers as "America's best and brightest" while also showing everyday Americans that the dream could belong to them, too. Simultaneously, the networks boosted their ratings as the "drama in space" drew one of the largest television audiences ever. It appeared to be a win-win situation for all involved.

Thus, it was in these four rhetorical strategies, including binary oppositions, that the ritual model of communication became evident on CBS, CNN, THC, and PBS. The reality of the Apollo project was created, repaired, transformed, and maintained through the mediated representations using nationalism, romanticism, pragmatism, and technology to reinforce American values in an effort to persuade the public to support Apollo and, more recently, NASA's manned missions into the final frontier. In more current broadcasts, particularly on CBS and CNN, the binary opposition has become the necessity of continuing space exploration to Mars, post-Apollo's zenith, despite not being able to afford the $100 billion price tag to do so.

While an analysis of how CBS, CNN, THC, and PBS presented the Apollo program in terms of ritual communication has been offered, there are several limitations. One is subjectivity—it was the author's own reading of the narratives presented on television. Other researchers, for example, might have extrapolated additional metaphors and interpreted different meanings in terms of their relevance to American values. A second limitation is that it was limited to one network, CBS, in the analysis of the Apollo years. Examination of an additional network such as ABC, for example, might have yielded a whole host of narratives generated from prized space reporter Jules Bergman. While it is true that networks pooled resources in the early days of space exploration and received the same video feed from NASA, each of the three networks in the mid- to late-1960s employed different anchors who explained the intricacies of the Moon shots. Indeed, the public tuned in to watch a favorite "personality"—he who was most trusted to get America to and from the Moon safely. Each of these three on-air personalities placed his own "spin" and interpretation on the success and the importance of the Apollo missions. Still today, after each broadcast, anchors offer comments to express their views regarding the topic at hand.

A similar but necessary limitation is that the analyses of later representations were limited to programs that specifically pertained to the Apollo program. With so many media outlets and the sheer number of advertisers who use clips of the famous Apollo footage, it would have been close to impossible to chronicle *all* of the mentions of Apollo. For example, future rhetorical and symbolic examination of Apollo narratives might attempt to explain the meaning of Volkswagen's advertisement (broadcast during the Apollo years) that pushed fuel efficiency as the "VW Bug" traveled across the Moon. Similarly, in the early 1980s, what might have been additional meanings of MTV's use of Aldrin saluting the American flag on the Moon as it began a new era of music catering to a culturally diverse "remote control" generation that would not only hear music but see it in action via the nascent music video? Today, how might computer-generated designs and the use of AI (artificial intelligence) depicting life on the Moon, based on Apollo missions' data collection, influence public opinion and environmental policy?

A fourth limitation is that audience response to the meanings expounded by broadcasters is not assessed. Although this is by design based on the Shoemaker and Reese argument that an overabundance of media research has focused on the audience rather than on content according to the media organization's intent, future analyses might take a phenomenological approach to gauge how audiences received narratives and metaphors used here. For instance, did audiences respond in the manner that NASA, CBS, and the other networks intended? What were and are the lived experiences of these viewers? To what narratives—romantic, frontier, patriotic, spiritual, pragmatic (what Carey called the uniting of community, the sharing of symbols)—did the public most relate in their decision to support or oppose Project Apollo?

Future studies might also compare differences in representations among print, broadcast, and digital media or compare the narratives that have been used to describe the differences between the Apollo and Shuttle programs in the news and in their remembrances, especially after the Challenger accident, when the news media ceased to accept NASA public relations claims and instead conducted their own investigations. A final suggestion might be to analyze the rhetoric used in special broadcasts or live coverage of the Apollo program. How, for example, did such rhetoric differ from the narratives in nightly news stories?

Since Apollo's last trip to the Moon in 1972, more than half of the world's population has been born. As we move farther away from those historic excursions, exegesis, especially of the American history teacher of choice—television and its streaming services in digital formats—becomes even more important to generations not yet alive during America's first venture to the Moon. As gatekeepers of culture, and as nations and private enterprises segue into the next chapter of space exploration, media, in all its different forms, will continue to construct realities.

NOTES

1. "A Small, Belated Step for Grammarians," *New York Times*, October 3, 2006, A19.
2. James Jeffrey, "Apollo 11: 'The Greatest Single Broadcast in Television History'," *BBC*, accessed July 10, 2019, https://www.bbc.com/news/world-us-canada-48857752.
3. Dorothy Nelkin, *Selling Science* (New York: W. H. Freeman and Company, 1987).
4. For examples, see Larry Williams, "Apollo 16 Coverage Fails to Soar," *The Commercial Appeal*, April 17, 1972. Gerald Nachman, "New Apollo Backup Team Logs Future Lunar Laughs," *New York Daily News*, April 28, 1972, 76. John Carmody, "A Shot at Space on PBS?" *The Washington Post*, November 10, 1972, B1. John Carmody, "Moonwalks: CPB Opts Out," *The Washington Post*, November 22, 1972, B1. John O'Connor, "Apollo 17 Coverage Gets Little Viewer Response," *The New York Times*, December 14, 1972, 87.
5. Herbert J. Gans, *Deciding What's News: A Study of CBS Evening News, NBC Nightly News, Newsweek, and Time* (New York: Random House, 1979), 50.
6. Michelle Zajac, "The Class of 1978 and the FLATS," NASA.gov, March 21, 2013, https://www.nasa.gov/history/the-class-of-1978-and-the-flats/. This class replaced the previous 1969 astronaut class.
7. James L. Kaufmann, *Selling Outer Space: Kennedy, the Media, and Funding for Project Apollo, 1961–1963* (Tuscaloosa, AL: The University of Alabama Press, 1994), 57.
8. Michael D. Lemonick, "Are We Losing Our Edge?" *Time*, February 13, 2006, 24.
9. Ibid.
10. Kim Krisberg, "Report: America Falling Behind in Global Science, Technology," *The Nation's Health* 54, no. 1 (February/March 2024): 5, https://www.thenationshealth.org/content/54/1/5.1.
11. Barbie Zelizer, *Covering the Body: The Kennedy Assassination, the Media, and the Shaping of Collective Memory* (Chicago, IL: The University of Chicago Press, 1992).

12. James Lee Kauffman, "Selling Space: The Kennedy Administration, the Media, and Congressional Funding for Project Apollo, 1961–1963" (PhD Dissertation, Indiana University, Bloomington, IN, 1989), 59.

13. Richard L. Campbell, "60 Minutes and the News: A Mythology for Middle America" (PhD Dissertation, Northwestern University, Evanston, IL, 1986), 34.

Appendix

Apollo Broadcasts on CBS, CNN, PBS, and THC

CBS COVERAGE OF APOLLO BY MISSION 1968–1972

Apollo Mission	Mission Dates	CBS Coverage Dates
7	10/11/68–10/22/68	9/4/68–10/25/68
8	12/21/68–12/27/68	10/28/68–2/20/68
9	3/3/69–3/13/69	2/25/69–5/12/69
10	5/18/69–5/26/69	5/13/60–5/28/69[1]
11	7/16/69–7/24/69	6/6/69–10/13/69[2]
12	11/14/60–11/24/69	10/20/69–12/18/69[3]
13	4/11/70–4/17/70	3/26/70–6/30/70[4]
14	1/31/71–2/9/71	11/2/70–3/1/71
15	7/26/71–8/7/71	3/12/71–10/18/71
16	4/16/72–4/27/72	1/5/72–5/2/72
17	12/7/72–12/19/72	12/4/72–12/22/72

CBS Coverage Post-Apollo, 1973–2022: 110 broadcasts, ranging from 20 seconds to 9 minutes.[5]

CNN Coverage Post-Apollo, 1995–2022: 33 broadcasts, ranging from 20 seconds to 31 minutes.[6]

PUBLIC BROADCASTING SYSTEM (PBS)

Original Broadcast Date	Name of Broadcast	Duration in Minutes
1/25/1978	*NOVA: One Small Step*	60
7/13/1999	*NOVA: To The Moon*	120
12/3/2014	*NOVA: First Man on the Moon*	53
12/26/2018	*NOVA: Apollo's Daring Mission*	54

THE HISTORY CHANNEL (THC)

Original Broadcast Date	Name of Broadcast	Duration in Minutes
6/14/1997	*20th Century with Mike Wallace: Crisis in Space: The Real Story of Apollo 13*	50
3/10/1999	*20th Century with Mike Wallace: Triumph & Tragedy: Apollo 11 and the Challenger*	50
1/22/2001	*Modern Marvels: Apollo 13*	50
8/24/2003	*History Sunday: Failure is Not an Option*	50
11/29/2003	*Special Presentation: Days That Shook the World: Wright Brothers/Moon Landing*	50
6/20/2004	*Tech Effect: Apollo 11*	50
7/21/2004	*Modern Marvels: Apollo 11*	50

NOTES

1. There was an overlap of stories as mentions of Apollo 10 began May 5 and the last one on June 2, 1969.

2. There were mentions, however, of the Apollo 11 mission before June 6, such as mentions of astronaut selection for the mission, for example, and mentions after October 13, including reports of scientific tests resulting from Moon rock sample analyses.

3. There was one mention prior to this date on September 8, 1969, that was included in the analysis.

4. There were two other stories pertaining to Apollo 13 outside of the March through June timeframe.

5. Using the key word "Apollo," for evening broadcasts only, there were 124 matches in the VTNA database, but 14 were unrelated to the space program.

6. Using key word "Apollo" for evening news broadcasts only, there were 38 matches in the VTNA database, but 5 were unrelated to the space program.

Bibliography

Aerospace Industries Association. "Industry Impact, 2023 Facts & Figures: U.S. Aerospace and Defense." Accessed May 15, 2024,https://www.aia-aerospace.org/news/aia-releases-2023-facts-figures-data-highlighting-ad-industrys-return-to-pre-pandemic-levels/.

Aldrin, Edwin E. Jr. "Lunar Dust Smelled Just Like Gunpowder." *Life*, August 22, 1969.

Aldrin, Edwin E. ("Buzz") and Warga, Wayne. "Return to Earth." *Good Housekeeping*, October 1973.

Allnutt, Robert F. Letter to the Honorable Goodell. NASA Headquarters, Washington, DC, January 16, 1968.

"And that's the way it is." *Government Executive*, January 1997.

Anderson, Steve. "History TV and Popular Memory." In *Television Histories: Shaping Collective Memory in the Media Age*, edited by Gary R. Edgerton and Peter C. Rollins, 21. Lexington, KY: University of Kentucky Press, 2001.

Banow, Joel. Interview with Martin Collins. Smithsonian Institution, National Air and Space Museum, Division of Space History, 17. February 1, 2000. National Air and Space Museum Archives, Accession number 2000-0027, Garber Archive Facility.

Benedict, Howard. "Full-Court Press: Apollo Meets the Media," *Air & Space,* June/July 1989, 83.

Bliss, Edward. *Now the News: The Story of Broadcast Journalism.* New York: Columbia University Press, 1991.

Bodnar, John. *Remaking America: Public Memory, Commemoration, and Patriotism in the Twentieth Century*. Princeton, NJ: Princeton University Press, 1992.

Bonney, Walter T. Memorandum to the NASA Administrator, NASA Public Information Program, August 20, 1959, NASA Headquarters History Office, Washington, DC.

Bonney, Walter T. Memorandum to the NASA Administrator, September 9, 1958. NASA Headquarters History Office, Washington, DC.

Bormann, Ernest G. "Fantasy and Rhetorical Vision: The Rhetorical Criticism of Social Reality." *Quarterly Journal of Speech, 58* (1972): 398.

Bowman, David. *This Must be the Place: The Adventures of Talking Heads in the Twentieth Century.* New York: HarperCollins Publishers, 2001.

Braunstein, Peter and Doyle, Michael William (Eds.). *Imagine Nation: The American counterculture of the 1960s and 70s.* New York: Routledge, 2002.

Buchwald, Art. "When it Comes to Moon Shots, Just Put Your Faith in Walter." *Washington Post*, May 29, 1969.

Bush, Alan J. and Boller, Gregory W. "Rethinking the Role of Television Advertising During Health Crises: A Rhetorical Analysis of the Federal AIDS Campaigns," *Journal of Advertising* 20 (1991): 31.

"Bush Sets 'New Course' for the Moon and Beyond," *NBC News*, January 13, 2004, https://www.nbcnews.com/id/wbna3950099.

Byrnes, Mark E. *Politics and Space: Image Making by NASA.* Westport, CT: Praeger Publishers, 1994.

Cage, John. Quoted in "Great Adventure." *Newsweek,* July 21, 1969.

Campbell, Richard. *60 Minutes and the New.* Urbana, IL: University of Illinois Press, 1991.

Carey, James W. *Communication as Culture.* New York: Routledge, 1989.

Carpenter, Ronald H. "America's Tragic Metaphor: Our Twentieth-Century Combatants as Frontiersmen." *Quarterly Journal of Speech* 9 (February 1990): 14.

Carpenter, Ronald H. "Frederick Jackson Turner and the Rhetorical Impact of the Frontier Thesis." *Quarterly Journal of Speech* 63 (April 1977): 117.

Carter, Jennifer Rudeseal. "View from the Bird Watch: Media, Memory, and America's Mercury Astronauts." PhD Dissertation, University of Southern Mississippi, Hattiesburg, MS, 1996.

CBS. Accessed February 21, 2024, http://www.cbs.com.

CBS Evening News, September 4, 1968 through December 31, 2022. Accessed via Vanderbilt Television News Archive, Nashville, TN.

CBS, *10:56:20 PM EST, 7/20/69: The Historic Conquest of the Moon as Reported to the American People by CBS News Over the CBS Television Network.* New York: Columbia Broadcasting System, 1970.

Chaikin, Andrew. *A Man on the Moon: The Voyages of the Apollo Astronauts.* New York: Penguin Books, 1994.

Christenson, Gale E. *Greenhouse: A 200-Year History of Global Warming.* New York: Penguin Books, 2000.

Clayton, James. "Apollo 11: Public Relations Dream." *Washington Post*, July 27, 1969.

CNN Evening News, January 1995 through December 31, 2022. Accessed via Vanderbilt Television News Archive, Nashville, TN.

Cohn, William. "History for the Masses: Television Portrays the Past." *Journal of Popular Culture* X, no. 2 (Fall 1976): 280–289.

Collin, Everett E. Interview by Rick Collin. NASA Aerospace Education Services Program Session Three, 31, August 6, 1992. NASA History Office Files, Record Number: 12664, Washington, DC.

Collins, Michael. *Carrying the Fire: An Astronaut's Journeys*. New York: Farrar, Straus, and Giroux, 1974.

Condit, Celeste. *The Meaning of the Gene: Public Debates about Human Heredity*. Madison, WI: The University of Wisconsin Press, 1999.

Cowan, Ruth Schwartz. *A Social History of American Technology*. New York: Oxford University Press, 1997.

Crawford, Ian. "To Still Boldly Go." *Prospect*, March 2003.

"Cronkite Feted by Scientists." *Denver Post*, November 22, 1971, 29.

Cronkite, Walter. *A Reporter's Life*. New York: Alfred A. Knopf, 1996.

Cronkite, Walter. "We are the Children of the Space Age: A Noted News Commentator Looks Back to Jules Verne and Forward to the Future." *TV Guide*, July 19, 1969.

Donovan, Robert J. and Scherer, Ray. "Live, from the Cape." *Air & Space,* October/November 1992, 68–71.

Dubrow, Dick. "Well, you've seen it: Buck Rogers is real." *The Evening Star*, July 21, 1969.

Eason, David L. "Telling Stories and Making Sense." *Journal of Popular Culture* 15 (1981): 125–129.

Edgerton, Gary R. "Television as Historian: A Different Kind of History Altogether." In *Television Histories: Shaping Collective Memory in the Media Age*, edited by Gary R. Edgerton and Peter C. Rollins, 10. Lexington, KY: University of Kentucky Press, 2001.

Ewen, Stuart. *PR!: A Social History of Spin*. New York: Basic Books, 1996.

Ferretti, Fred. "Cronkite on Endurance: 'You Don't Think of That.'" *New York Times,* July 24, 1969.

Fogo, Fred. "'Come Together, Over Me:' Generational Memory and Social Drama in the Death of John Lennon." PhD Dissertation, The University of Utah, Salt Lake City, UT, 1990.

Gans, Herbert J. *Deciding What's News: A Study of CBS Evening News, NBC Nightly News, Newsweek and Time*. New York: Vintage Books, 1979.

George Mason University. "Center for History and New Media." Accessed February 4, 2004, http://chnm.edu/tools/surveys/details/70.

Glenn, John. *P.S. I Listened to Your Heart Beat: Letters to John Glenn*. Houston, TX: World Book Encyclopedia Science Service, Inc., 1964.

Goldman, Steven. "Images of Technology in Popular Films: Discussion and Filmography." *Science, Technology & Human Values* 14, no. 3 (Summer 1989): 275–301. https://doi.org/10.1177/016224398901400305.

Gottfried, Jeffrey. "Americans' Social Media Use." Pew Research, 2024. https://www.pewresearch.org/internet/2024/01/31/americans-social-media-use/.

"Great Adventure." *Newsweek*, July 21, 1969, 73.

Gregory, Jane and Miller, Steve. *Science in Public: Communication, Culture, and Credibility*. New York: Plenum Trade, 1998.

Haines, Harry. "Mediated Vietnam: The Politics of Postwar Representations." PhD Dissertation, The University of Utah, Salt Lake City, UT, 1987.

Halbwachs, Maurice. *The Collective Memory*. New York: Harper Colophon, 1950.

Halbwachs, Maurice. *On Collective Memory*. Translated by Lewis A. Coser, ed. Chicago, IL: University of Chicago Press, 1992.

Haney, Paul. *Newsweek*, July 28, 1969, 27.

Harris, Gordon L. *$elling Uncle Sam*. Hicksville, NY: Exposition Press, Inc., 1976.

Hartley, John. *Understanding News*. London: Metheun Publishing, Ltd., 1982.

Hasian, Marouf and Plec, Emily. "The Cultural, Legal, and Scientific Arguments in the Human Genome Diversity Debate." *The Howard Journal of Communications* 13 (2002): 304.

Henderson, Artis. "To the Moon and Back." *Gulfshore Life*, July 30, 2019. https://www.gulfshorelife.com/2019/07/30/to-the-moon-and-back/.

Henige, David P. *Oral Historiography*. New York: Longman, Inc., 1982.

Hewitt, Don. *Tell Me a Story: Fifty Years and 60 Minutes in Television*. New York: Perseus Books Group, 2001.

Hoopes, James. *Oral History: An Introduction for Students*. Chapel Hill, NC: University of North Carolina Press, 1979.

The History Channel (THC). 20th Century with Mike Wallace, *Crisis in Space: The Real Story of Apollo 13*. Originally aired June 14, 1997.

The History Channel (THC). 20th Century with Mike Wallace, *Triumph & Tragedy: Apollo 11 and the Challenger*. Originally aired March 10, 1999.

The History Channel (THC). History Sunday, *Failure is Not an Option*. Originally aired August 24, 2003.

The History Channel (THC). Modern Marvels, *Apollo 11*. Originally aired July 21, 2004.

The History Channel (THC). Modern Marvels, *Apollo 13*. Originally aired January 22, 2001.

The History Channel (THC). Special Presentation, *Days that Shook the World: Wright Brothers/Moon Landing*. Originally aired November 29, 2003.

The History Channel (THC). Tech Effect, *Apollo 11*. Originally aired June 20, 2004.

History of Manned Space Flight, NASA Publication #75-24641. Washington, DC: National Aeronautics and Space Administration, 1975.

Hogan, A. R. "Walter Leland Cronkite." *Ad Astra*, July/August 1991.

Hurt, III, Harry. *For All Mankind*. New York: The Atlantic Monthly Press, 1988.

James, Doug. *Walter Cronkite: His Life and Times*. Brentwood, TN: JMPress, 1991.

Jeffrey, James. "Apollo 11: 'The Greatest Single Broadcast in Television History.'" *BBC*, Accessed July 10, 2019. https://www.bbc.com/news/world-us-canada-48857752.

John F. Kennedy Presidential Library and Museum. "Address at Rice University on the Nation's Space Effort, September 12, 1962." https://www.jfklibrary.org/learn/about-jfk/historic-speeches/address-at-rice-university-on-the-nations-space-effort.

John F. Kennedy Presidential Library and Museum. "Address to Joint Session of Congress, May 25, 1961." https://www.jfklibrary.org/learn/about-jfk/historic-speeches/address-to-joint-session-of-congress-may-25-1961.

Jones, Thelma. Letter to Julian Scheer, NASA Chief PAO. March 18, 1969, NASA History Office.

Kaplan, Judith and Muniz, Robert. *Space Patches: From Mercury to the Space Shuttle.* New York: Sterling Publishing Co., Inc., 1986.

Kaufmann, James. *Selling Outer Space: Kennedy, the Media, and Funding for Project Apollo, 1961-1963.* Tuscaloosa, AL: The University of Alabama Press, 1994.

Kauffman, James. "Selling Space: The Kennedy Administration, the Media, and Congressional Funding for Project Apollo, 1961-1963." PhD Dissertation, Indiana University, Bloomington, IN, 1989.

Kauffman, James. "A Successful Failure: NASA's Crisis Communications Regarding Apollo 13." *Public Relations Review* 27 (Winter 2001): 437–438. https://doi.org/10.1016/S0363-8111(01)00099-6.

Kitch, Carolyn L. *Pages from the Past: History and Memory in American Magazines.* Chapel Hill, NC: The University of North Carolina Press, 2005.

Kitchell, James. Quoted in "Remembering the Space Race: A Panel Discussion" held at the National Air and Space Museum of the Smithsonian Institution in Washington, DC on November 15, 2000. Transcribed in *Quest, The History of Spaceflight Quarterly*, v. 9, n. 3 (2002).

Krisberg, Kim. "Report: America Falling Behind in Global Science, Technology." *The Nation's Health* 54, no. 1 (February/March 2024): 5. https://www.thenationshealth.org/content/54/1/5.1.

Krug, Linda. *Presidential Perspectives on Space Exploration: Guiding Metaphors from Eisenhower to Bush.* New York: Praeger, 1991.

Lantry, Douglas N. "Man in the Machine: Apollo-Era Space Suits as Artifacts of Technology and Culture." *Winterthur Portfolio* 30, no. 4 (1995): 220.

Lattimer, Richard L. *All We Did Was Fly to the Moon.* Gainesville, FL: Whispering Eagle Press, 1985.

Launius, Roger D. *Apollo: A Retrospective Analysis, Monographs in Aerospace History 3.* Washington, DC: National Aeronautics and Space Administration, 1994.

Launius, Roger D. "Heroes in a Vacuum: The Apollo Astronaut as Cultural Icon." *Florida Historical Quarterly* 87, no. 2 (2020): Article 6.

Launius, Roger D. "Perceptions of Apollo: Myth, Nostalgia, Memory, or all of the Above?" *Space Policy* 21 (2005): 130–131.

Launius, Roger D. "Tech Effect: Apollo 11." *The History Channel*, aired June 20, 2004.

Launius, Roger D. and Hunley, J. D. *An Annotated Bibliography of the Apollo Program, Monographs in Aerospace History 2.* Washington, DC: National Aeronautics and Space Administration, 1994.

Laurent, Lawrence. "Moon Mission Ratings Race." *Washington Post*, July 25, 1969.

Lemonick, Michael D. "Are We Losing Our Edge?" *Time*, February 13, 2006.

Levi-Strauss, Claude. *The Raw and the Cooked: Introduction to a Science of Mythology*, Vol. I. Translated by John and Doreen Weightman. Chicago, IL: University of Chicago Press, 1969, 1983.

Lyotard, Jean-Francoise. *The Postmodern Condition: A Report on Knowledge.* Minneapolis, MN: The University of Minnesota Press, 1984.

MacKenzie, Donald. *Inventing Accuracy: A Historical Sociology of Nuclear Missile Guidance.* Cambridge, MA: MIT Press, 1990.

Maher, Neil. "Shooting the Moon." *Environmental History* 9 (July 2004): 528.

Malone, Bill C. *Country Music, U.S.A.* Austin, TX: University of Texas Press, 1985.

Malone, Bill C. *Don't Get above Your Raisin': Country Music and the Southern Working Class.* Urbana, IL: University of Illinois Press, 2002.

McCurdle, Dorothy. "The U.S. Flag from the Moon Voyage Furled at $25,000." *Washington Post*, June 14, 1970.

McCurdy, Howard E. *Space and the American Imagination.* Washington, DC: Smithsonian Institution Press, 1997.

Meyrowitz, Joshua. *No Sense of Place: The Impact of Electronic Media on Social Behavior.* New York: Oxford University Press, 1985.

Middleton, David and Edwards, Derek. *Collective Remembering.* London: Sage Publications, 1990.

"The Moon and Middle America," *Time*, August 1, 1969, 11.

Morris, Steven. "How Blacks View Mankind's 'Giant Leap.'" *Ebony*, September 1970.

MSN.com. "History of the Nike Swoosh and the Birth of a Multi-Billion Dollar Brand." *The Street.* Accessed June 27, 2024. https://www.msn.com/en-us/money/companies/history-of-the-nike-swoosh-and-the-birth-of-a-multi-billion-dollar-brand/vi-BB1nl7Py?item=themed_featuredapps_enabled?loadin=defaultbrowser#details.

Murray, Charles and Cox, Catherine Bly. *Apollo: The Race to the Moon.* New York: Simon & Schuster, 1989.

Museum TV. "Cronkite Remembers." Accessed June 12, 2024. https://museumtv.pastperfectonline.com/archive/7438F209-966D-4BE2-A497-163975494915.

NASA. "The 1980s: All Eyes Focus on Space Shuttle," Accessed June 30, 2024. https://www.nasa.gov/history/the-1980s-all-eyes-focus-on-space-shuttle/.

NASA. "The Apollo 1 Tragedy." Accessed June 10, 2024. https://nssdc.gsfc.nasa.gov/planetary/lunar/apollo1info.html.

NASA. "Functions of the Administration." *Space Act of 1958.* Accessed May 1, 2024. https://www.nasa.gov/history/national-aeronautics-and-space-act-of-1958-unamended/

NASA. Letter to John Houbolt. NASA History Office, Washington, DC.

NASA. *NASA Facts: "Spacemobile,"* August 22, 1991. NASA History Office, NASA Headquarters, Washington, DC. File Record Number: 12664.

NASA. *NASA, Space: The New Frontier.* Washington, DC: GPO, 1963.

NASA. News Release. 04-229, July 16, 2004. NASA History Office.

NASA. "The Outer Space Treaty of 1967." Accessed June 18, 2024. https://www.nasa.gov/history/SP-4225/documentation/cooperation/treaty.htm.

NASA Press Kit Release No: 68-208. Apollo 8, December 15, 1968, NASA Headquarters, Washington, DC.

NASA Press Kit Release No: 69-83K. Apollo 11, July 6, 1969. NASA Headquarters, Washington, DC.

NASA. "Race to Space." Accessed June 20, 2024. https://www.nasa.gov/stem-content/race-to-space/.

NASA. *The Vision for Space Exploration.* NASA Printing Office, Washington, DC, February 2004.

"NASA's Public Information Policy." *The Washington Post*, March 27, 1971.

NBC Commentary. "Why is the Moon So Boring." *The Observer,* May 6, 1972.

Nelkin, Dorothy. *Selling Science: How the Press Covers Science and Technology.* New York: W.H. Freeman and Company, 1987.

Newcomb, Horace M. and Hirsch, Paul M. "Television as a Cultural Forum: Implications for Research." In *Interpreting Television: Current Research Perspectives*, edited by Willard D. Rowland, Jr. and Bruce Watkins. Beverly Hills, CA: Sage Publications, 1984, 58–73.

Nishimoto, Warren. "Interviewing in Oral History and Journalism." *Oral History Recorder,* Summer 1992, 2.

Obensen, Tambay. "'Chasing the Moon': Watch Civil Rights Protest at 1969 Moon Launch in Rare Clip – Exclusive." *IndieWire*. Accessed June 20, 2024, https://www.indiewire.com/features/general/chasing-the-moon-pbs-ralph-abernathy-1202153386/.

O'Connor, John E., ed. *American History, American Television: Interpreting the Video Past.* New York: Frederick Ungar Publishing Co., 1983.

O'Connor, John J. "TV: Schirra and Cronkite again Team for Apollo 15." *New York Times*, August 4, 1971.

Paine, Thomas O. Memorandum to Associate and Assistant Administrators and Center Directors, February 25, 1969. NASA History Office, NASA Headquarters, Washington, DC.

Parry, Pam. *Eisenhower: The Public Relations President.* Lanham, MD: Lexington Books, 2016.

PBS. NOVA, *Apollo's Daring Mission.* Originally aired December 26, 2018.

PBS. NOVA, *First Man on the Moon*. Originally aired December 3, 2014.

PBS. NOVA, *One Small Step*. Originally aired January 25, 1978.

PBS. NOVA, *To the Moon.* Originally aired July 13, 1999.

PBS.org. "Meet the People: Walter Cronkite." Accessed June 19, 2024. https://www.pbs.org/avoidingarmageddon/meetThePeople/meet_03_03_01_cronkite.html.

Platoff, Anne M. "Where No Flag Has Gone Before: Political and Technical Aspects of Placing a Flag on the Moon." NASA Contractor Report 188251. NASA History Office, Washington, DC.

"'Put an American Flag on the Moon' – Bennett." *Space Daily*, June 3, 1969.

Pynchon, Thomas. *Gravity's Rainbow*. New York: The Viking Press, Inc., 1973.

"Remembering the Space Race: A Panel Discussion." National Air and Space Museum of the Smithsonian Institution in Washington, DC, November 15, 2000. Transcribed in *Quest, The History of Spaceflight Quarterly*, v. 9, n. 3 (2002).

Robinson, Gertrude Joch. "Television News and the Claim to Facticity." In *Interpreting Television: Current Research Perspectives*, edited by Willard D. Rowland, Jr. and Bruce Watkins. Beverly Hills, CA: Sage Publications, 1984, 199–221.

Roland, Alex. "Late night with Aaron Brown." *CNN News*, aired January 14, 2004.

Rotary National Award for Space Achievement. "Corona Award." Accessed June 19, 2024, https://www.rnasa.org/other.html.

Rushing, Janice Hocker. "Mythic Evolution of 'The New Frontier' in Mass Mediated Rhetoric." *Studies in Mass Communication* 3 (1986): 265–296.

Sadtler, Helen Cruikshank. "Flags on the Moon." NASA Headquarters, Washington, DC.

Schudson, Michael. *Watergate in American Memory: What we Remember, Forget, and Reconstruct about the Past.* New York: Basic Books, 1992.

Shoemaker, Pamela and Reese, Stephen. *Mediating the Message: Theories of Influences on Mass Media Content*, 2nd ed. White Plains, NY: Longman Publishers, 1991.

Silverstone, Roger. *The Message of Television: Myth and Narrative in Contemporary Culture.* London: Heinemann, 1981.

"A Small, Belated Step for Grammarians." *New York Times,* October 3, 2006.

"Space History." *Space.com.* Accessed January 9, 2004. http://www.space.com/news/spacehistory/scheer_obit_040901.html.

Spigel, Lynn. "Outer Space and Inner Cities: African American Responses to NASA." In *Welcome to the Dreamhouse: Popular Media and Postwar Suburbs,* edited by Lynn Spigel. Durham, NC: Duke University Press, 2001, 141–182.

Spigel, Lynn. "White Flight." In *The Revolution Wasn't Televised: Sixties Television and Social Conflict*, edited by in Lynn Spigel and Michael Curtin. New York: Routledge, 1997, 47–72.

Stark, Steven D. *Glued to the Set: The 60 Television Shows and Events that Made Us Who We are Today.* New York: The Free Press, 1997.

Streeter, Thomas. "An Alternative Approach to Television Research: Developments in British Cultural Studies at Birmingham." In *Interpreting Television: Current Research Perspectives*, edited by Willard D. Rowland, Jr. and Bruce Watkins. Beverly Hills, CA: Sage Publications, 1984, 74–97.

Tonkin, Elizabeth. *Narrating Our Pasts: The Social Construction of Oral History.* Cambridge: Cambridge University Press, 1992.

United States Congress, Senate. *Committee on Aeronautical and Space Sciences, Subcommittee on NASA Authorization.* May 2, 1960.

United States Congress, Senate. *Committee on Aeronautical and Space Sciences, Subcommittee on NASA Authorization, NASA Authorization for Fiscal Year 1961, Part 1, 86th Congress, 2nd Session*, 29. March 30, 1960. https://www.nasa.gov/history/SP-4002/p1a.htm.

University of California, Santa Barbara. "Richard Nixon Inaugural Address." The American Presidency Project. Accessed June 18, 2024. https://www.presidency.ucsb.edu/documents/inaugural-address-1.

Utley, Garrick. *You Should Have Been There Yesterday: A Life in Television News.* New York: Public Affairs, 2000.

Van Dyke, Vernon. *Pride and Power: The Rationale of the Space Program.* Urbana, IL: University of Illinois Press, 1964.

Watson, Mary Ann. *The Expanding Vista: American Television in the Kennedy Years.* New York: Oxford University Press, 1990.

Weitekamp, Margaret. *Right Stuff, Wrong Sex: America's First Women in Space Program.* Baltimore, MD: The John Hopkins University Press, 2004.

White, Theodore. "Camelot, Sad Camelot." *Time*, July 3, 1978.

The White House. "President Bush Announces New Vision for Space Exploration Program," January 14, 2004. Accessed March 23, 2024. https://georgewbush-whitehouse.archives.gov/news/releases/2004/01/20040114-3.html.

Wilcox, Dennis L. and Camerson, Glen T. *Public Relations: Strategies and Tactics*, 8th ed. Boston, MA: Pearson/Allyn and Bacon, 2006.

Williams, Raymond. *Television and Technological Form*. London: Collins and Sons, 1974.

Zajac, Michelle. "The Class of 1978 and the FLATS." *NASA.gov*, March 21, 2013. Accessed December 14, 2023. https://www.nasa.gov/history/the-class-of-1978-and-the-flats/.

Zelizer, Barbie. *Covering the Body: The Kennedy Assassination, the Media, and the Shaping of Collective Memory*. Chicago, IL: The University of Chicago Press, 1992.

Zimmerman, Robert. *Genesis: The Story of Apollo 8*. New York: Dell, 1999.

Index

About the Author

Kathy K. Previs, PhD, is a professor of communication at Eastern Kentucky University and an online instructor at West Virginia University. A former NASA employee, her research interests focus on rhetorical strategies in public relations (PR) campaigns pertaining to science and health, PR in sports and tourism, and ethical practices in PR. Previs's work has been published in *Science Communication*, *Public Relations Review*, *Communication Research*, and *International Journal of Public Opinion Research*, to name a few. Previs holds a BS in communication from Virginia Tech, a MS in mass communication from Middle Tennessee State University, a PhD in mass communication from Ohio University, and has earned certificates in history from OU's Contemporary History Institute and in plant-based nutrition and health from Cornell University. She taught at Charleston (SC) Southern, Lipscomb, and Vanderbilt Universities before joining the PR faculty at EKU. When she's not enjoying time with her husband and beloved rescue and AKC champion Dalmatians, you'll find her searching for a new bicycle trail.

www.ingramcontent.com/pod-product-compliance
Lightning Source LLC
LaVergne TN
LVHW091530200926
848483LV00002B/348
* 9 7 8 1 6 6 6 9 7 4 9 0 4 *